L. N Badenoch

True Tales of the Insects

L. N Badenoch
True Tales of the Insects
ISBN/EAN: 9783337072902

Printed in Europe, USA, Canada, Australia, Japan
Cover: Foto ©berggeist007 / pixelio.de

More available books at **www.hansebooks.com**

TRUE TALES

OF

THE INSECTS

b

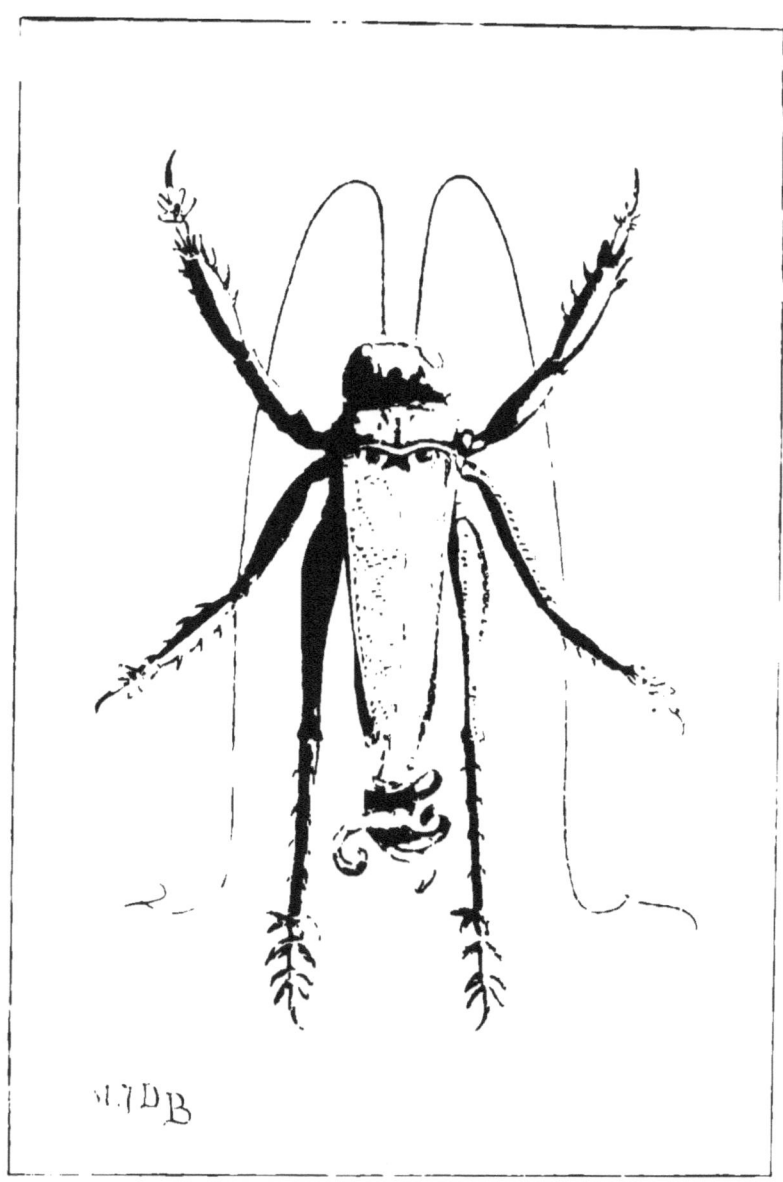

SCHIZODACTYLUS MONSTROSUS.

TRUE TALES

OF

THE INSECTS

BY

L. N. BADENOCH

AUTHOR OF "ROMANCE OF THE INSECT WORLD"

WITH FORTY-FOUR ILLUSTRATIONS BY MARGARET J. D. BADENOCH

NEW YORK: E. P. DUTTON & CO.
LONDON : CHAPMAN & HALL, LD.
1899

PRINTED BY
WILLIAM CLOWES AND SONS, LIMITED,
LONDON AND BECCLES.

INSCRIBED TO THE
DEAR MEMORY OF D . . .

PREFACE.

A FEW of the Essays in this volume have been published in serials. "Symbols of Psyche" originally appeared as an article in *The Sunday at Home*. "Day-flying Moths" and "The Case Moths," but for some slight alterations, and some additions, have appeared in *Knowledge* and *Appletons' Popular Science Monthly*, and I have to thank the editors of those periodicals for kind permission to reproduce them here.

To Mr. W. F. Kirby and Mr. C. O. Waterhouse, of the British Museum (Natural History), I must express my thanks for the kindness with which they have always helped me with the literature of the subject, and have given me the free use of the valuable collections under their charge.

LONDON,
December, 1898.

CONTENTS.

ORTHOPTERA.
CURSORIA.

CHAPTER I.
THE DEVIL'S RIDING-HORSE (MANTIDÆ).

"Devotional" attitude—Legends and superstitions arising therefrom—Meaning of habit—The tiger, not the saint of the insect world—Not only carnivorous, but pugnacious, and a cannibal—Voracity not limited to insects—Modification of front legs—Their principal function—Typical development of an insect's leg—Modification in detail of leg of Mantis—The limb in repose—Its secondary functions—The other legs—Modification of prothorax—Its remarkable elongation and mobility—Development shows importance—Not legs and thorax alone, whole organization in conformity with a carnivorous life : head, sense-organs, etc., organs of flight—Atrophy of wings—Coptopteryx females 3

CHAPTER II.
THE DEVIL'S RIDING-HORSE (MANTIDÆ)—*continued.*

Egg-laying and the egg-capsules—Capsule of *Mantis religiosa ;* situation of eggs, subsidiary parts, consistency, explanation of manner of formation—Capsules of other Mantidæ—Parasites—Metamorphoses of Mantis—Emergence, appearance, and interesting life of young—Development of organs of flight—Nymphs compared with sub-apterous or apterous adults—Latter with apterous Phasmidæ—Protective and Aggressive resemblance—Modifications of forms in general, of the Ground Species, of the Plant Types   Special protective resemblance—Alluring colouring—Aggressive mimicry – Geographical distribution 16

CHAPTER III.

WALKING-STICKS AND WALKING-LEAVES (PHASMIDÆ).

General peculiarities—Appearance grotesque—Mesothorax often relatively very large—Yet tegmina usually of small size, or absent, even where lower wings are very largely developed—In such cases provision for defence of latter organs essential – Characteristics and habits—Are herbivorous—Their immobility—Reasons and use thereof—More means of defence : prickles and spines, power of ejecting nauseous fluid, aquatic habits—Curious power of reproducing lost or injured limbs—Eggs in Phasmidæ generally of a most remarkable nature—Specially exemplified in eggs of Phyllium—Change during scramble out of egg ; and after—Change in colour at different periods of life 41

CHAPTER IV.

WALKING-STICKS AND WALKING-LEAVES PHASMIDÆ —continued.

Marvellous imitative resemblance of vegetative objects—Walking-sticks proper—Beautiful and giant winged forms—Bizarre shapes galore—End gained by this mimicry of course protection against attack—There is perhaps no other group of insects which in form and colour are so generally imitative—Leaf-insects—Resemblance to leaves displayed by tegmina, also by other parts—Female alone possessed of large leaf-like tegmina—Tegmen of female Phyllium from various points of view an exceptional structure—Success of artifice demonstrated—Distribution of family—That Walking-sticks come of a remote antiquity—That they are a singularly isolated group 60

SALTATORIA.

CHAPTER V.

LOCUSTS AND GRASSHOPPERS (ACRIDIIDÆ).

Series Cursoria and Saltatoria—To two families of latter, term Grasshopper applied—Acridiidæ, prominent characteristics—Anatomy—Organs of special sense—Air-sacs—Acridiidæ remarkable amongst Orthoptera for possession of—Arrangement in Rocky Mountain Locust—Use—Locust an aëronaut—To this fact largely due its enormous powers of flight—Dilatable tracheæ—Locust, how endowed intellectually—Gift of "song"—Of great importance to the creatures—Apparatus for producing sound—The music, characteristic of male—Stridulation during flight—Acridian ears—Three forms—Minute structure—Ears in both sexes, as in most species—Function of acoustic organs difficult to determine—Possible solution of difficulty—Oviposition; and philosophy of egg-mass—Migratory locusts may make a deposit of eggs at more than one place during migration—Egg-enemies—Process of escape of the young from the egg, of much interest—Post-embryonic development of Acridiidæ—Change of colour in course of development ... 81

CHAPTER VI.

LOCUSTS AND GRASSHOPPERS (ACRIDIIDÆ)—*continued.*

Most species of Acridiidæ not migratory—Species ascertained to be migratory—Migratory disposition not caused by anatomical differences: migratory species exist in countries without giving rise to swarms—Ravaging power of migratory locusts; huge size of swarms—Famine and pestilence probable sequences—No law of periodicity governing destructive flights—Phenomenon of migration explained by excessive multiplication—Other causes, both immediate and remote—Remarkable manifestations of instinct attend migration—Disappearance of locusts from a spot invaded apparently obscure: they again migrate after growth to land of ancestors—"Voetgangers," interesting points in their natural history—How these wingless locusts cross rivers—Distance to which swarms may migrate—Length of a single flight—Proof of power of prolonged flight that they are able to cross large bodies of water—Natural enemies, vertebrate and invertebrate—Some species of Acridiidæ present an unusual aspect—Oedipodides, striking cases of colour difference—Is correlative with locality—Interesting Eremobiens; modified to extraordinary extent for desert life ... 106

CHAPTER VII.

GREEN GRASSHOPPERS (LOCUSTIDÆ).

Distinguishing characters—Absence of air-vesicles connected with tracheal system—Ocelli generally imperfect—Peculiar head-ornament—Locusta, ovipositor—Life-histories—Deposition of eggs in galls, of *Meconema varium*, of *Xiphidium ensiferum*—Development of *Microcentrum retinerve* — Tibial ears, their structure and functions—Musical organs and music—Katydids, their pertinacity — As pets—Ill-effects of confinement—Food-habits—Anabrus, increase to large numbers—Perfection of resemblance of tegmina to leaves—Defence of a positive nature—Resemblance to Stick Insects—Cave-dwellers—Remarkable forms—Gigantic and ugly Anostostoma—Of the curious genus Deinacrida—*Schizodactylus monstrosus* - Eumegalodonidæ 141

LEPIDOPTERA.

CHAPTER VIII.

SYMBOLS OF PSYCHE.

Fable of Psyche—Its origin—In its later aspects—Conspicuous beauty and abundance of the Symbols—Favourite resorts—Not only gregarious, but migratory—Twilight fliers—Of a quarrelsome disposition—Peculiarities of highest interest—Different females of *Papilio memnon* — Perfect protective imitative resemblance of Kallima, the leaf-butterflies 163

CHAPTER IX.

DAY-FLYING MOTHS.

Lepidoptera divided into Rhopalocera (butterflies, or day-flyers) and Heterocera (moths, or night-flyers)—Antennæ unsatisfactory as a classificatory basis : no one set of characters will serve as an infallible guide to distinguish moths from butterflies—"The series of affinities in nature a concatenation or continuous series"—This more or less gradual blending a strong argument in favour of community of descent—Castniidæ in some respects combine the characters of both Lepidopterous divisions—Have most affinities with moths—*Megathymus yuccæ*, the Yucca Borer, an interesting aberrant form—Regarded by some as a genuine butterfly—Habitat, appearance, habits—Depredations committed : white powdery bloom : funnel-like tube, characteristic of larva—Brilliant and graceful Uraniidæ proved to belong to the Heterocera—*Urania boisduvalii*, one of the most beautiful Lepidoptera known ; and others—Migratory habits of day-flying moths 184

CHAPTER X.

THE CASE MOTHS (PSYCHIDÆ).

In structure of female, and in habit, the strangest and most abnormal of all Lepidoptera—Females have become degenerate—Young sack-bearers at home—Ingenuity in construction of case—*Metura elongata* a most interesting architect—Lictor Moth—*Animula huebneri*—Curious case of *Animula herrichii*—Enlargement, repair, locomotion, temporary suspension, complete withdrawal, mode of moving and retaining position within case—Coming of Age of male—Supposed assistance of long sticks at event—After-life, and general characteristics—Want of homogeneousness in perfect state of insects of this group—Probable cause of disappearance of beauty of male—Singular rarity of moths considering abundance of cases—Females perpetual prisoners, living and dying within larva habitaculum 203

CHAPTER XI.

THE HAWK MOTHS SPHINGIDÆ.

Leading characteristics of this favourite group—Posterior spine or caudal horn—Sphinx-like attitude—Highly specialized condition of some of the structural characters, of peculiar interest—Macroglossinæ — Chœrocampinæ — Ambulicinæ — Sphinginæ — Manducinæ —Smerinthinæ—Sphingidæ have their metropolis in Tropics—Occasional visitors from sunny climes ... 219

CHAPTER XII.

THE DEATH'S HEAD MOTH.

In several respects a most remarkable species—"Grim feature"—It has a voice—Many theories put forward to account for the cry—Sound emitted by pupa—By larva; its nature and cause—Moth of superior dimensions, of nocturnal habit—Not surprising it should be an object of alarm to the superstitious—As a bee-robber—That its stridulous voice controls the bees—It is excessively sluggish—At sea—That it is nomadic in habit—Times of appearance in imago; in larva—Single- or double-brooded with us—On rearing the Death's Head Moth 230

INDEX 245

LIST OF ILLUSTRATIONS.

NO.			PAGE
	SCHIZODACTYLUS MONSTROSUS ...	*Frontispiece*	
1.	HARPAX TRICOLOR AND PSEUDOCREOBOTRA...	*Headpiece*	3
2.	MANTIS RELIGIOSA IN DEVOTIONAL ATTITUDE	...	7
3.	*A.* FORE-LEG OF MANTIS RELIGIOSA; *B.* LEG OF A BEETLE (LUCANUS CERVUS)		10
4.	A STICK-LIKE MANTIS (LEPTOCOLA GRACILIMA) WITH ATROPHIED WINGS		11
5.	HEAD AND PROTHORAX OF MANTIS RELIGIOSA, WITH INSERTION OF ANTERIOR LEGS		13
6.	EGG-CAPSULES OF MANTIDÆ	*Headpiece*	16
7.	EGG-CAPSULE OF MANTIS RELIGIOSA		19
8.	A GROUND MANTIS (EREMIAPHILA TYPHON), FROM EGYPT ...		26
9.	CILERADODIS RHOMBICOLLIS, WITH PROTHORACIC EXPANSION AND LEAF-LIKE ELYTRA		27
10.	DEROPLATYS TRUNCATA, FROM BORNEO		32
11.	GONGYLUS TRACHELOPHYLLUS, WITH NUMEROUS FOLIACEOUS LOBES		33
12.	TROPIDODERUS RHODOMUS ...		47
13.	EGGS OF DIFFERENT WALKING-STICKS		53
14.	ACROPHYLLA TITAN	*Headpiece*	60
15.	A STICK INSECT (PHANOCLES CURVIPES), FROM ST. VINCENT		61
16.	CEROYS LACINIATUS, FROM NICARAGUA: IRREGULAR LEAF-LIKE EXPANSIONS PROTRUDE ALL OVER IT		68
17.	GRAEFFEA COCCOPHAGUS		74
18.	A WALKING-LEAF (PHYLLIUM SCYTHE), FROM SILHET		75

No.			PAGE
19. A Grasshopper, Truxalis Pharaonis		*Headpiece*	81
20. Teratodes Monticollis ...	... ...	...	93
21. Proscopia Inæqualis, which bears a Great General Resemblance to a Stick Insect		...	107
22. A very Aberrant and Beautiful Grasshopper (Pneumora Scutellaris) ...	... ...	...	119
23. Methone Anderssoni, specially Remarkable for its Complex Organs for Producing Sound ...		...	131
24. Pterochroza Ocellata, its Tegmina Resembling Leaves			142
25. A Cave-dweller (Dolichopoda Palpata)		... ...	143
26. Eumegalodon Ensifer, one of the most Remarkable of the Locustidæ, from Java	... ...	...	151
27. Symbols of Psyche, etc. ...		*Headpiece*	163
28. The Dragon's Bride ...	... ...	...	165
29. Hetaira Esmeralda, from Brazil. A Clear-wing Butterfly ...	... ...	...	169
30. The Calliper Butterfly (Charaxes Kadenii), from Java: sucking Liquid from a Muddy Spot ...		...	170
31. Morpho Menelaus, from Tropical America. Brilliant Metallic Blue and Black	...	...	171
32. Cheritra Jaffra; Brown with White Tails; from Burmah			178
33. Different Females of the Malayan Papilio Memnon		...	179
34. A Leaf-butterfly Kallima Inachis, in Flight and in Repose ...	... ...	...	182
35. Urania Braziliensis; Migratory; from Brazil.		*Headpiece*	184
36. Castnia Eudesmia, from Chili ...	... ...	...	189
37. Yucca Borer (Megathymus Yuccæ, in Flight and in Repose; from the United States	... ..	...	191
38. Urania Boisduvalii; Green and Velvety Black; from Cuba ...	... ...		196
39. Larva Case (Metura Elongata), from Sydney			207
40. Male and Female Metura Elongata ...			213
41. A Long Proboscis (Cocytius Cluentius) ...	...	...	221
42. Hawk Moth (Lophostethus Dumolinii, from Port Natal.			225
43. The Death's Head Moth		*Headpiece*	230

ORTHOPTERA

TRUE TALES OF THE INSECTS.

CURSORIA.

CHAPTER I.

THE DEVIL'S RIDING-HORSE (MANTIDÆ).

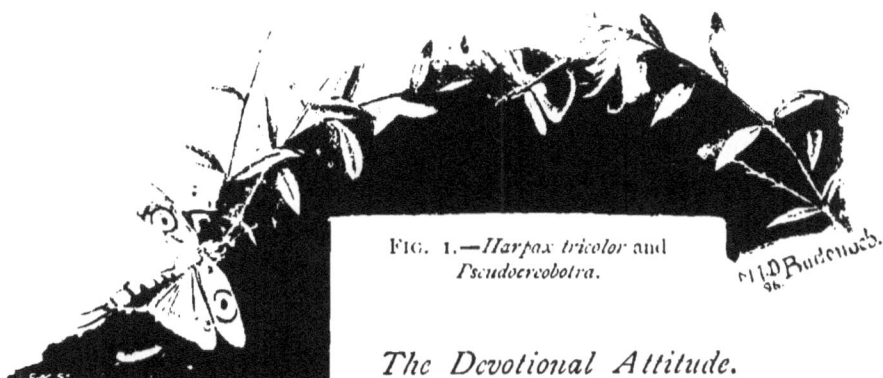

FIG. 1.—*Harpax tricolor* and *Pseudocreobotra*.

The Devotional Attitude.

PROBABLY no other insects are surrounded with such a halo of legend and superstition, none have acquired the same reputation for wisdom and saintliness, as those technically known as the Mantidæ. The character they bear is one of long standing, and almost world-wide fame. It originates in the insects' habitual attitude, that appears devotional; but appearances are risky things to judge by. For hours they rest motionless upon the four hind limbs in the characteristic posture, with the head raised upon the elongated and semi-erect prothorax, and with the front legs entirely free, half-opened, the joints clasped together, held like uplifted hands in prayer. To our illogical and superstitious forefathers, what could this position denote if not devotion?

The name Mantis (diviner, soothsayer, seer, prophet), it is said, was bestowed on the insect by the Greeks, in accordance with the idea that, when in its normally motionless pose, it is engaged in meditation on futurity. Amongst the Turks and Arabs it is of a quasi-sacred nature, and they hold that it prays with its face towards Mecca. One again meets with the notion of its peculiar habit as an act of invocation or piety in the name of Prega-Diou, or Prie-Dieu, as it is called by the Provençals, and in Italy; in Portugal, it is the Louva Dios; while the English-speaking nations, somewhat clumsily, dub it the praying insect; and many more or less similar common names—preacher, nun, saint, suppliant, mendicant—applied to it in Southern Europe, testify to the general reverence with which it is regarded. In Languedoc and other provinces, where it is abundant, as indeed at whatever place the superstition prevails, it is deemed a crime to injure it, and at least most culpable neglect not to remove it out of harm's way, should it appear exposed to the slightest possibility of danger. If a Hottentot by accident kill or maim the local species, he is believed to be thereby doomed to ill-luck for life, and never afterwards can shoot buffalo or elephant.

In fact, in Africa, both among the Hottentots and among certain tribes to the north, this strange feeling of veneration attains its highest limits, amounting, as some allege, to actual worship. In the case of the former people, should one of these insects chance to alight on an individual, he immediately becomes a saint in their eyes, a special favourite of Heaven.

Monkish legends go the length of making the mantis give utterance to its devout sentiments; a specimen settling on the hand of St. Francis Xavier, he desired it to sing the praises of God, whereupon it carolled forth a very beautiful canticle. "So divine a creature is this esteemed," says Mouffet, "that if a childe aske the way to such a place, she will stretch out one of her feet, and show him the right way, and seldom or never misse."

Not Saints, but Tigers.

But in one corner of the world, at any rate, its outward actions seem to be taken for what they are worth, as indicated by the Brazilian, somewhat uncomplimentary title that heads this chapter. In reality a more atrocious little savage could not be found—it lives by rapine; its tastes are essentially carnivorous. It is as observant and quick as a monkey, as sly and stealthy as a cat: it is the tiger, not the saint, of the insect world. Its so-called devotional attitude (see Fig. 2) is simply nothing but a lying in wait for what the gods may send in the shape of food. Establishing itself, as if in rapture, upon some twig or leaf, it will remain thus absolutely stationary, prepared to seize any unwary insect that may fall within convenient reach. After it exhibits a wonderful degree of patience, let us say an insect happens to alight within a short distance of it. Instantly it catches sight of the newcomer, and begins, with slow, silent tread, to steal towards it. So imperceptible the motion, it can only be appreciated by steady and prolonged watching. At the same time the fore legs, which up till now had been bent back upon themselves, commence to open. Little by little, the hunter creeps near its unconscious prey, its goggle eyes staring upon this object of absorbing interest. At last it is close enough to strike; and, with celerity of movement the eye cannot follow, a formidable foreleg is shot out to its full length, and brings back the victim, hopelessly secured and crushed between the shank and thigh, and scarcely more than a moment is lost ere the body is torn to pieces and devoured. Again the mantis assumes its ecstatic mood, in readiness, if need be, for the insidious progress which is part of its crooked policy. No insect, however agile, can escape those merciless paws, the rapidity of the stroke, just described, of which is in marked contrast to this being's other movements, which are of some slowness.

Although ants seem exempt from approach, mosquitos,

FIG. 2.—*Mantis religiosa* in devotional attitude.

flies, and small bees feed the insatiable appetite of the mantidæ, they destroy caterpillars and large grasshoppers in great numbers, and make war on walking-sticks and beetles; but the tougher morsels are generally discarded for more succulent delicacies; and some kinds of insects, the Meloës in particular, are evidently extremely repulsive to them on account of their secretion. These carnivorous habits by no means figure alone as the sole enormity in the private character of our insects. They are of a most quarrelsome disposition, and cannibalism trips up the heels of their pugnacity. From their very birth the larvæ fight. If two or more adults be shut up together, they engage in a desperate conflict, cutting at each other with their sword-like legs, until one of the belligerents falls in the fray, when the conqueror swallows up his antagonist; the male, being the smaller, often constitutes the feast. Their manœuvres while joining battle have been likened to those of hussars with sabres, and sometimes one cleaves the head of the other from the body with a single slash.

Aware of the pugnacious propensities, the Chinese indulge their talents for gambling by feeding them and keeping them apart in little bamboo cages, and matching them like fighting cocks, laying wagers on the results. The mantidæ do not limit their voracity to insects; they vanquish creatures which, from their size and strength, one would have thought were totally free from their attacks. Large South American species seize, and eat small frogs, lizards, and even birds. Surely little vertebrates, taken by surprise and feeling themselves pressed in the terrible arms, are at once so overcome by terror as to be incapable of offering resistance.

The Organization is in Conformity with a Carnivorous Life.

By now, doubtless, the fact of the important part played by the fore legs in the carnivorous habit will have presented itself forcibly to the reader's notice. It is they that serve to seize living prey, and the form these organs have taken for this purpose renders them the most characteristic feature of the mantidæ. An insect's leg, it may be stated, is composed of four chief parts— the coxa, femur, tibia, and tarsus; or the hip, thigh, shank, and foot—and its typical development may be seen by referring to Fig. 3, *B*. But in the mantidæ, the front pair of limbs suffers modification, principally in the direction of increased strength and size (see Fig. 3, *A*). The coxa is elongated, and becomes slender and prismatic, and its articulation, remarkably mobile, is such that the whole limb has a great freedom of action, and is used, much as an arm might be used, in conveying food to the mouth. The third joint or thigh is robust and compressed, and bears on its curved under side a channel, furnished on each edge with strong, movable spines; ordinarily stronger, fewer, and farther apart at the external edge than on the inner border, where they are numerous and crowded: this armed channel extends only along the terminal half of the femur, and towards the base of the armature there is one or several long spines in the centre of the thigh. Into the groove, in repose, the tibia fits, it being shorter than the femur, horny, and compressed, and terminated by a long, sharp, curved claw. Its under side is equally occupied by a double row of fixed spines or teeth. This stout piece, bending back on the femur, forms with the latter a veritable hand, as well as a powerful vice; for the two rows of spines of the tibia and those of the femur oppose each other, and work into each other, the former chiefly with those on the inner border of the thigh; the spines on the external border of the thigh perform the duty rather of a kind of boundary, to prevent the tibia swerving outwards. It is clear, that in the act of grasping it is the inner border that is best armed by far, as it needs be, since, naturally, it is on this side that the insect devours its prey. The slender tarsus does not call for special remark; it is united with the tibia at the base of the claw, and can be completely effaced by applying itself in a wonderful manner against the inferior side of the femur.

FIG. 3.—*A*. Foreleg of *Mantis religiosa*: *B*. Leg of a Beetle (*Lucanus cervus*).

In repose, be it observed, the three principal portions which go to make up the leg are all folded back the one upon the other, and obtain concealment beneath the prothorax; the coxæ, in contact with each other, being enclosed between the under side of the prothorax and the femora.

The front legs may likewise assist their owners in walking, as the existence of normal tarsi betokens, and the sharp claw of the tibiæ may even be useful in climbing the trunks of trees, but the principal function of these limbs is as a powerful weapon for the capture of prey.

The intermediate and posterior pairs of legs are without distinction — slender, generally long, and fitted only for walking; normally cylindrical, they sometimes are provided with membranous lobes of variable form.

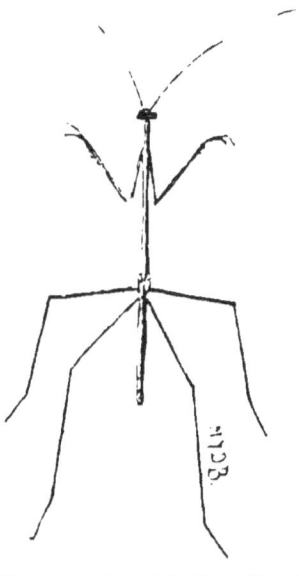

FIG. 4.—A stick-like Mantis (*Leptocola gracilima*) with atrophied wings.

Next to the enlarged front legs, with their adaptation to a particular end, the most striking lineament of these insects is the long prothorax, it being the longest segment of the body; among slender types (see Fig. 4) the lengthening may become marvellously pronounced. This piece is, in fact, elongated into a narrow neck, rather dilated in front, above the insertion of the anterior legs, to give added strength to their articulation (see Fig. 5), leaving a long, flat, narrow space on the under side behind these legs; the remainder of the thorax is short, and hid by the wings and wing-covers in the position of rest.

This remarkable development shows the importance of this part of the body, the movements of which play no mean *rôle* in the pursuit of prey. Its articulation with the meso-thorax is very supple; it is raised obliquely, it may be turned to either side with the greatest liberty, the body meanwhile being supported solely by the two posterior pairs of legs. The lengthening of the pro-thorax, we may believe, has been promoted by the habitual projection in front of the rapacious legs, and to their presence towards the end of the long elevated neck is in part due the mobility of its mesothoracic articulation.

Not that the legs and thorax are the only parts arranged with reference to the wants of the insects. Their whole organization, as with all animals that give chase to living prey, is superior, and is in direct conformity with a carnivorous life, which demands at the same time perfected senses, and great vigour and suppleness. The head, instead of being bound in the prothorax, enjoys great mobility, and may be turned in every way, even so as to bring the mouth obliquely upwards. Its customary position is vertical, or bent up against the prothorax, the mouth applied against the neck. It is triangular—sometimes rounded or long— especially with the great bulged eyes, which in general occupy the superior angles, to enable the insects to see in many directions. The size of the visual organs, no less than their position, denotes a well-developed sense; and simple eyes, or ocelli, are never wanting. The body, though sometimes suffi- ciently thickset, is usually long and narrow.

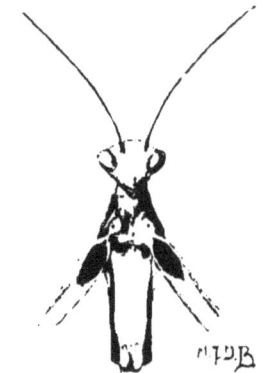

FIG. 5.—Head and prothorax of *Mantis religiosa*, with insertion of anterior legs.

As one would expect among hunting insects, the organs of flight are often amply developed, and powerful enough to cover considerable distances. The upper pair, or elytra as they are called, are as long as the abdomen, or sometimes extend in both sexes beyond the extremity of that piece, over which they are carried horizontally when closed, one greatly overlapping the other. Compared with the hinder pair, they are at least as long as they, with a few exceptions, but they never become enormously abbreviated, and always serve as sheaths or covers to the wings. They are coriaceous or membranous, those of the males constantly longer and narrower than in the females, and always more transparent. The wings are large, and, being well protected in rest, remain membranous, though they often extend a little beyond the elytra; the exposed tip tends to become indurated.

Atrophy of the Wings.

There exist frequent cases, however, where the organs are atrophied more or less, especially among the females. The abbreviation is pushed the farthest in mantidæ of stick-like form, for a body excessively lank and long almost precludes the power of flight, and condemns the species which possess it to a pedestrian life (see Fig. 4). To these insects flight would be possible only with wings of great size, and the narrowness of the body does not permit the muscles to enlarge sufficiently to move such organs easily; whereas the bigger mantidæ of this shape have wings always more or less undeveloped.

But the females seem never entirely apterous; on the thorax always persists some trace of the organs of flight. Thus among certain species (Coptopteryx) where the elytra alone become developed to a rudimentary state, even in these rare instances the wings are not altogether obliterated, but are invariably indicated under the form of striate lobes, as in the nymphs.

In addition to the previously mentioned characters, the males have longer antennæ than the females, they have larger eyes, a narrower prothorax, and longer and more slender raptorial legs; but more decided sexual differences occur.

CHAPTER II.

THE DEVIL'S RIDING-HORSE (MANTIDÆ)—*continued.*

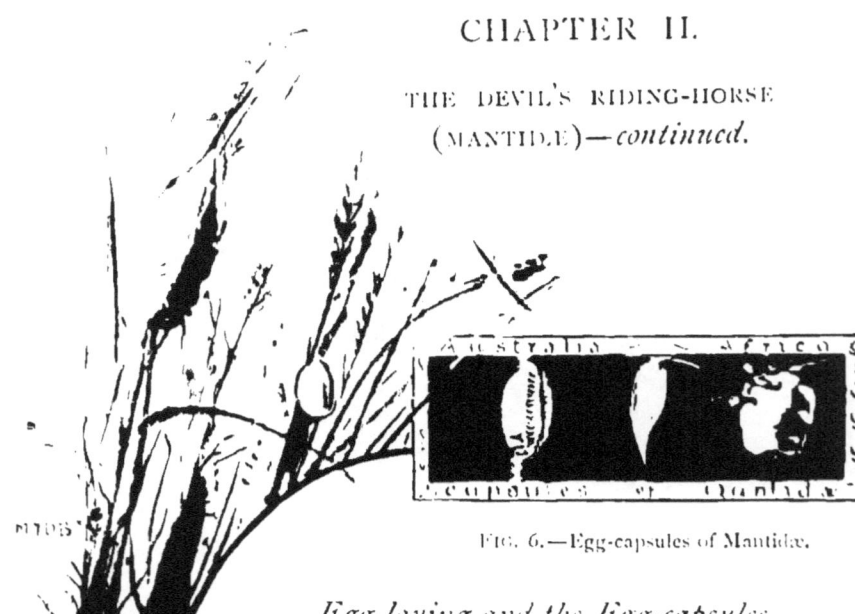

FIG. 6.—Egg-capsules of Mantidæ.

Egg-laying and the Egg-capsules.

IN Europe the egg-laying of the mantidæ occurs in September. Among the kinds observed, the insect discharges at the same time as the eggs, a mass of viscous or gummy matter, which she affixes to the stems of bushes or on stones, and, as it dries in the air, it assumes the form of a fair-sized spherical or oval capsule or case, and surrounds and shelters the eggs. The whole mass terminates in a sort of neck that sticks to the stem, or is directed upwards, lending the capsule the appearance of a seed or fruit, for which indeed it has been frequently mistaken. These capsules, or oöthecæ, as they are technically called, vary with the species (see Fig. 6). Many are most fragile, some of much beauty, and their internal structure is of marvellous regularity; it is no exaggeration to say that, at the first glance, one might easily take them to be organized bodies. If generally the parent dies after laying, this is not always the case, for a mantis has been known to fashion successively four capsules, and even to establish six different ones, at intervals of seven or eight days; but in Europe the parent invariably succumbs before the arrival of cold weather. Seeing, then, that the young brood produced in autumn remain in the egg state until the following summer, the mantidæ disappear entirely during about six months, for there is only one annual laying. But in warm climates, in all probability, the life of these insects does not suffer an interruption equally great, and already, in southern parts of Europe, we find indication of the fact in the genus Empusa, which hibernate in the state of larva, and transform in the subsequent spring.

An examination of the ellipsoid capsule of the common mantis, *Mantis religiosa*—by the way, it attains a length of more than three centimetres—is well worth making (see Fig. 7). The gummy mass voided by the insect has been spread out, we see, into a succession of layers, fitting lightly by their curvature, rounded and equalized on the surface in the form of the ovoid cap, the layers being of insignificant thickness. A transverse cutting of the capsule shows that each layer is divided into three parts; a median chamber or sac, like a flat bottle, open at the top, containing, towards the bottom, the eggs, which are narrow, longish, yellow in colour, and to the number of eight to ten, resting by their large end, and adhering to the floor of the mass; and all ranged on the same plan, somewhat fan-wise, diverging towards the floor, and in most symmetrical order, one half on either side, with or without a central egg; moreover, each is enveloped in a gummy pellicle. The two walls of this flat bottle or cell which contains the eggs are very tough and chitinous, and each partition narrows above the space enclosing the eggs, and terminates in a neck, by an arched lamina or scale, imbricating with the lamina of the partition above, in the direction of the small end of the mass. Together these laminæ constitute a scaly band on the median line of the convex upper side of the capsule, dividing the surface into two symmetrical parts. By the terminal scale of each partition imbricating with the scale of the partition following, the cell with the eggs placed between the two partitions is closed, that is to say, a sort of elastic operculum is formed, which, after hatching, the larvæ have to raise in effecting their egress through the neck.

Laterally to the important compartment with the eggs, on either side are envelopes, formed of a foamy substance, chitinous, but very light, made up also of arched cells, and disposed by successive layers, corresponding to the succession of the central chambers. To these they must be regarded as mere protective casings, since they contain no eggs.

FIG. 7.—Egg-capsule of *Mantis religiosa*.

It is also in the light of protection that one must look upon the layers of the two extremities of the capsule. They do not contain eggs, and consequently present no horny central cell, but are composed solely of cellular tissue, analogous to that of the two lateral zones of the other layers; they simply afford additional strength to the ends of the structure. The highest and lowest cells that envelop eggs are smaller than the rest, and possess only two, four, or six of these little objects, for whose well-being alone they were established.

The capsule is at first soft and whitish, but soon darkens, and becomes firm, parchment-like, or very resistant, and tears only with difficulty, and is impermeable to water. It may be plunged into liquid without the eggs being injured, so admirable the closure of the opercular scales one upon the other.

It is not easy to understand how the mantis, simply by the act of laying, contrives to form a structure so wonderfully regular, and so complicated as this capsule; it would be a clever bit of work were it built by the mouth and feet, but fashioned as it is, it appears a conjuror's trick. The insect begins to establish the edifice at the large end; and whilst the viscous matter flows, the abdomen is caused to assume a circular, undulatory motion, ceaselessly working up the gummy mass, and arranging it in the successive concave layers. It must be supposed that for each layer she at first discharges the eggs clothed, as it were, in gummy stuff, which in drying hardens, and becomes the central horny sac, and then to right and left she deposits a viscid fluid less strong, known to us eventually in its solidified state in the shape of the lateral foamy cells.

Curiously enough, the eggs at the small end of the oötheca hatch first, although these were the last deposited.

The European Ameles, small mantidæ which likewise inhabit the Mediterranean basin, have prismatic capsules, about two centimetres long, composed of a most neat series of triangular cells, each enclosing six or seven eggs, almost destitute of lateral cellular tissue.

In spite of all the mother's care, chalcidideous parasites infest her eggs to some extent. Several species have been obtained from the eggs of mantidæ of Mauritius and Brazil, and those of *Mantis religiosa*, in Cephalonia, are known to be attacked. It is probable that the parasite seizes her opportunity to start the career of her progeny before the glutinous covering has acquired its final consistency.

Development of the Young.

Step by step, we may study the development of *Mantis religiosa*—which may be taken as the type of the mantidæ—and see these eggs, on the whole securely housed, change to larva, to pupa, and at last to imago. To the lot of the mature insect short life falls; in turn it propagates its kind, and that done, in submission to the winter's cold, it dies. The eggs do not hatch till June, and only after a lapse of nearly three months the mantis arrives at the perfect state.

The young, in quitting the egg, leaves the shell at the bottom of the cell. It already presents, in a general way, the aspect of the adult. Too feeble to use its legs, it employs a special artifice in escaping from the cell where it is confined. On the chitinous covering of the body spines are developed, directed backwards, and these acting as a prop against the walls of the cell, the larva, by giving an undulatory movement to the abdomen, is enabled to make its way towards the operculum with ease. There, with its back, it raises the opercular scale which closes it in, and gains its liberty to the outer world in the most natural manner.

No sooner is the flap pushed the least bit ajar than it shuts to by its own elasticity. It acts, indeed, like the spring lid of a box, so that the posterior legs and the long anal threads of the larva may be caught as in a vice; and should the spring be too strong, and the larva be unable to profit by the assistance in drawing itself from its skin, it perishes, for want of power to extricate itself. In its other sheddings the insect is no less obliged, in order to get rid of the useless tight case, to fix to some object, in default of which it has no option but to work itself out of what has become its prison, by tearing its coat with its claws.

A single capsule gives birth to from fifty to a hundred larvæ.

These creatures, in form almost the very image of their parents, are in the beginning excessively delicate, soft, and pale, and upon issuing from their cradle they disperse, but do not venture far from its immediate vicinity. No long time is spent ere they betake themselves to hiding, among leaves and under stones, there to undergo their first moult. As yet they have not partaken of food, nevertheless they already exhibit the suppleness of the adults, and the same poses, turning their head backwards, and holding the anterior legs folded up upon themselves, as in the actual preparatory attitude of attack. In spite of this bearing, they are singularly timid, and dart under stones at the approach of an insect. Their anterior legs still serve merely for climbing; they constantly, in fact, creep to the extremities of twigs, where they appear to repair in search of prey. During their early days their pains seem all for nought, for the sight of plant-lice, for which they are on the lookout, causes them the greatest perturbation, and it is only with extreme circumspection, and by taking as many precautionary measures as prudence can dictate, that the larvæ dare to draw near, and, little by little, acquire the trick of seizing them. The young mantidæ now commence to have the capability of partaking of a fairly varied insect diet. At the end of twelve or fifteen days they experience a second moult—a fatal crisis for many, for those which cannot manage to cast off the now too small envelope, die in the endeavour. By this time the mantis has advanced in its killing powers, to be able to master insects as large, for instance, as the Ephemeræ. About a fortnight or three weeks later, it must again strip off its skin. In all, there are at least seven moults, but it is impossible so speak on this point with absolute certainty, owing to the difficulty of rearing these larvæ in captivity.

After each moult the mantis is languid and sickly, and unequal to the chase, and relapses into the timorous state of the period of extreme youth, so that but a glimpse of an insect near by suffices to throw it into a condition of violent agitation and terror; but soon the pangs of a voracious appetite cast fear and timidity to the winds, and it freely attacks and subdues to its needs other larvæ and insects. By-and-by its agility becomes such, that it not only climbs on tree-trunks with remarkable ease in pursuit of prey, and passes from branch to branch with the utmost facility, but it takes on the movements of the monkey which we meet with among adults, letting itself fall from one branch to another, hanging suspended, and recovering itself by aid of the long, sharp tibial claw.

The organs of flight appear under the form of simple prolongations of the teguments of the lateral borders of the segments of the meso- and metathorax; and in these stumps can be distinguished some of the principal parts, and they hold the normal position. The last moult suddenly develops the elytra and wings in all their extent, an enormous and truly astonishing development, one still unexplained, when we compare the voluminous organs with the little sheaths in which they were contained among the nymphs.

It is the nymphs that approach the sub-apterous adult mantidæ, as among Coptopteryx females, in which the wings, although not separated from the metathorax, exist in the state of stumps, under the form they assume during the nymph period. But whilst among mantidæ, in the case of the non-separation of the organs of flight, it is always the nymph state that persists, among Phasmidæ, on the contrary, the larva state is that which is preserved most often : in the last-mentioned Orthoptera, when the wings are wanting, they are entirely obliterated. The development of the mantidæ, then, is arrested at a later stage, a fact indicating a more advanced step in transformation, and a nature more perfect.

Protective and Aggressive Resemblance.

A most interesting feature of the mantidæ is their presentment, in a high degree, of a phenomenon which indeed the whole order Orthoptera presents—that of adaptation to their conditions of life. In a more or less wonderful manner, their appearance harmonizes with their surroundings, with the soil or vegetation on which they live, tending to their concealment, and so escape from enemies, when the resemblance is protective ; and the same circumstance is of special value in enabling them to lie in wait for, and to creep upon their prey, in which cases the resemblance is aggressive. This one character—that of adaptation leading to concealment, then—is of use to its possessors for two different ends, for defence and for attack.

Mantidæ invariably imitate the colour of the spots which they inhabit, and as the greater number of species live on plants and shrubs, they have a green colour. Green is almost the normal case among mantidæ. An exception to the green kinds that habitually repose upon vegetation is comprised by those which, imitating dead leaves, take a brown colour.

But there is another exception to this rule of colour among the plant-types. Sometimes *Mantis religiosa* is grey. This deviation may be probably regarded as due merely to the influence of the sun in arid places, where vegetation is itself parched and scanty; but it appears to become hereditary in the spots where it is oftenest produced, and where foliage is well-nigh wanting. In sandy

FIG. 8.—A ground Mantis (*Eremiaphila typhon*), from Egypt.

and rocky desert districts it occurs, and thus a race is developed which, in assimilating itself with the general artistic effects or colour of its environment, passes unobserved, without attracting the fatal attention of enemies. Not only this, but in such districts the individuals of the green colour, being most conspicuous, will

FIG. 9.—*Choeradodis rhombicollis*, with prothoracic expansion and leaf-like elytra.

be readily devoured; and generation after generation the grey form will tend to repeat itself, to the gradual exclusion of the other; so also the grey form will disappear in verdant regions, where the green specimens will be better protected, and will propagate their kind. In this way, certain species tend to divide into two varieties, the one of a green colour appertaining to vegetation, the other grey, living amid sands and rocks.

There are species which live exclusively on sterile rocks and plains, where the alteration of colour has become an accomplished fact; they have always a grey or yellow tint, in conformity with that of the soil on which they reside. Those insects which assume colours other than green may be looked upon as a younger form, replacing an older type. They belong almost entirely to the Old World; the species of the New World appertain essentially to the class of green mantidæ.

Those Curious Creatures the Eremiaphilæ.

As an example of the ground-types, take those curious creatures (Eremiaphila, see Fig. 8) first discovered by Savigny at the time of the 1798 expedition to Egypt. Not by their colour alone, but by the rugosity of their body, these insects imitate the earth. They are essentially dwellers in deserts; they inhabit deserts deprived of all vegetation, and enjoy power of adaptation to their surroundings such that they always offer the most perfect identity with the shade of the sands and pebbles on which they move, giving rise to local varieties. Lefebvre, who described them much later than Savigny, mentions that he was unable to discover any insect capable of nourishing these carnivora, and they disappeared at the borders of oases, as soon as vegetation commenced ; they were confined to the absolutely arid sands where they were met. However, species have since been found in spots that support a few sorry plants. The genus Eremiaphila seems localized in the Mediterranean regions of Africa and Asia, appearing in Egypt, the desert of Luxor, the oasis, isthmus of Suez, in Abyssinia, Nubia, Algeria, and Arabia.

As to the general modifications of form, one may say of the ground mantidæ, and among the Eremiaphilæ in particular, the body tends to become clumsy and squat, with the anterior legs short and large, and the organs of flight atrophied. The plant-types, on the contrary, tend to lengthen : those with a slender neck and large abdomen live on leaves; the stick-like body of others renders them in keeping with nothing so much as with slender herbaceous or woody stems ; their wings tend to abbreviate, and even become unfit for purposes of flight in the case of the larger species of this form, as Thespis ; small kinds of the type have wings well developed.

Beautiful Examples of Special Resemblance.

It is the tropical plant-types that present in a wonderfully perfect degree the phenomenon of Protective and Aggressive resemblance, not by colour merely, reproducing the general effect of the surroundings; the resemblance may be strikingly special, in which the appearance of some particular vegetative object is more or less exactly copied in colour, and also in outline and shape. Certain mantidæ have the veining of their wings modified so as precisely to imitate that of a leaf. The body in mantidæ, no less than that in walking-sticks, is subject to carry appendages, giving rise to most bizarre forms, which imitate parts of plants. But the resemblance is not produced by the same contrivances as among Phasmidæ. Here it is in general the prothorax that dilates in the form of a leaf (*Chœradodis, Epaphrodita*); or the elytra have a cut-out outline (*Deroplatys*), or expand beyond measure (*Stagmatoptera, Cardioptera*), producing a marvellous resemblance to great leaves with their nervures. It is hardly possible to conceive vegetable appearances more grotesque than those brought about by the development of a number of foliaceous lobes with which the legs, the body, and head are adorned (*Gongylus*); or when the insects even associate certain postures with their appendages, so as to resemble follicular fruits, or packets of leaflets (*Acanthops*).

But as compared with those of the walking-sticks, all these adventitious lobes are less indented; they are plainer and more defined, and the body is never covered with spines, properly so called.

At a glance one may recognize the genus Chæradodis, by the great membrane which extends from each side, and occupies the length of the long prothorax (see Fig. 9). These insects are present not only in tropical America, as in Costa Rica, Guayaquil, New Granada, Ecuador, and Brazil, but in India, including Ceylon; and the flattened shape of *Chæradodis rhombicollis*, its colour a delicate green, in part of a pale red almost yellow, the large flat, rhomboidal prothoracic dilation, with the lateral angles rounded, the long narrow opaque green leaf-life elytra, the transparent wings, the somewhat denticulate anterior legs,—these features all combine to make an insect as odd as it is interesting.

The genus Deroplatys, which is apparently exclusively Asiatic, replacing in the ancient world the genus Acanthops, is divided into two kinds, in virtue of its strange leaf-like prothoracic appendage. It may be large towards the front and small behind, in which case the form is often quite grotesque; or the reverse, it attains the greatest size at the posterior end, and is more or less triangular, but the shape often differs in the two sexes. *Deroplatys truncata* (see Fig. 10), a species that must be relegated to the latter class, is to be met with in

Singapore, at the extremity of the Malay Peninsula, in Borneo, and Java; *dessicata* belonging to the first class, in Malacca and Java; and Sumatra is another island that gives birth to these mantidæ. Their colour dried is a greenish yellow and brown feuillemort. Among the females especially, the elytra have a variable form, being sometimes large, truncated, and rounded, sometimes longer and pointed, the edges, as we have seen, often irregular, while the wings, coloured and frequently ornamented with beautiful spots and arched bands, may suddenly narrow into a kind of little tail, reaching beyond the elytra in repose, as in *truncata*. The thighs of the second and third pair of legs carry at

FIG. 10.—*Deroplatys truncata*, from Borneo.

FIG. 11.—*Gongylus trachelophyllus*, with numerous foliaceous lobes.

their extremity foliaceous dilatations, and the abdomen is dilated and possessed of small projections. These insects, in the form and structure of the elytra, by the lobes on the sides of the abdomen, and their colour like a faded leaf, and in other ways, approach near to the Acanthops.

In the genus Gongylus the prothorax is wonderfully elongated, as it were into a slender stem, but dilated in leaf form above the anterior legs ; the head terminates in a double leaflet ; the elytra widen abruptly at the base—they are longer and demi-membranous among the males, opaque among the females; the thighs of the two posterior pairs of legs carry at their extremity three rounded lobes; and the abdomen is expanded (see Fig. 11).

Alluring Colouration and Aggressive Mimicry.

The Special Aggressive Resemblance of mantidæ assumes yet another phase—the disguise is used for more than concealment, and does more than hide the insect from its prey ; it may even serve as a direct means of securing the latter, attracting them by simulating some object which is to them of particular value. Such appearances, which observation has failed to discover in many cases besides the Mantidæ in the realm of insects, Dr. Wallace calls Alluring Colouration, and they constitute some of the most curious and interesting forms of aggressive resemblance.

One of the most beautiful and remarkable examples is that of a wingless Indian mantis (*Hymenopus bicornis*), of great rarity, which, both in colour and form, resembles an orchid, or some similarly fantastic flower. The whole insect is brilliantly pink. Its large and oval abdomen represents the labellum of an orchid, and the thighs of the four posterior legs are immensely dilated and flattened into broad, pear-shaped plates, the apparent petals of a blossom; so that when seated motionless, as is its wont, amid bright green foliage, with thorax and abdomen raised at right angles to one another, with the forelegs drawn out of sight under the thorax, and the four expanded thighs of the other legs spread out two on each side, it is conspicuous, of course, but presents a complete and deceptive imitation of a gay-hued flower. Here colour, form, and attitude all conspire, in an inimitable manner, to produce the resemblance. Of the meaning of the resemblance there can be no doubt. Insects seem attracted to the mantis, as insects to flowers; they settle upon it, and are instantly captured.

A very similar species, which, when at rest, lying in wait for its prey, exactly resembles a pink orchid, inhabits Java. This mantis is said to feed especially on butterflies, so that by its imitation of a flower, the insect it feeds on will actually be attracted towards it; in fact, "it is really a living trap, and forms its own bait."

Equally interesting instances of the striking simulation to flowers are exhibited by Indian Mantidæ of the genus Gongylus, the floral resemblance, by deceiving and attracting insects, serving to secure for the pupal mantidæ a supply of food. With regard to their under surface, the leaf-like prothoracic expansion is coloured either white, or a pale bluish-violet, inclining to mauve, and acquiring a reddish tinge towards the margins, so as to resemble a flower with a white or a purple corolla, and both species have the same blackish-brown blotch in the centre, thus resembling the opening of a tube in the middle of the corolla of a flower. A specimen having a bright violet-blue thoracic shield was found in Pegu, by the late Mr. S. Kurz, and its resemblance to a flower for a moment deceived the practised eyes of the botanist.

The resemblance of mantidæ to the excreta of birds is also of use to the insects for aggressive purposes, since flies are known to be attracted by such droppings. One mantis closely resembles the white ants on which it feeds—an instance of the somewhat rare phenomenon of Aggressive Mimicry.

Geographical Distribution.

The mantidæ appertain essentially to the warm countries of the world, being especially abundant in the tropics, and becoming fewer towards high latitudes; they do not go beyond the temperate regions, and never penetrate to the cold parts; to speak more precisely, as a whole they hardly pass the 46 of latitude in the southern hemisphere, the 48° to the north, so that they do not occur in England. In North America they seem to stop in Pennsylvania, and there are few species in the Southern States; in South America they reach the confines of Patagonia. In Europe, so far as the central portion is concerned, while they are usually arrested on the northern slope of the Alps, this is not invariably the case, for during the last century *Mantis religiosa* was common to the environs of Ratisbon, though to-day it might be looked for there in vain. Several species are met with in the south of France, and the mildness of climate in Western Europe permits of the insects spreading along the coast-line to Normandy.

Being neither travellers nor vagrants, seas, generally speaking, seem to offer an insuperable barrier to their advance, and the different forms that serve as types may be said to remain attached to the regions where they are developed; the geographical distribution of these insects, in fact, is distinguished for this trait—the tendency towards localization of the genera to the different continents. In this way the distribution is very clearly defined. Thus, in America the genera belong for the most part exclusively to the New World, and differ from those of the Old World; and in the latter, by the side of widespread genera, others are peculiar to Africa, or to Asia, or to Australia, and to the Isles of the Pacific. Of the genera represented in the two hemispheres, their number is sufficiently restricted that it may be said they are exceptions to the common rule. There are the genera Ameles and Iris, found both in America and in the Mediterranean regions; Liturgousa and Cardioptera appear in America and in Southern Africa; Miopteryx in America and Asia; and lastly, the American Thespis crop up in Africa and in Asia in the shape of Oxythespis. It will be remarked that while Ameles and Iris are of the Old World types, which may have passed to America, Liturgousa, Cardioptera, and Thespis have a real American stamp, and seem rather to have emigrated from America to Africa, at an epoch more or less remote.

Distinctly, America appears to possess fewer genera and species than the Oriental hemisphere, and it is in Asia, particularly in the Moluccas and the Isles of Sunda, that the family seems to be most richly represented.

Mantis religiosa, which has been often mentioned, a sufficiently pretty, though not striking mantis, is full of interest to us, as being the species best known near home, across the Channel, to our neighbours in France. In all the south of that country it is common, but is rarer in the north, reaching Saintes, La Rochelle, Dijon, and the coast of Normandy as far as Havre; and it has been taken occasionally near Paris, at Fontainebleau, etc. It occurs in Spain, Italy, and Sicily, frequently in the fields and gardens of Tuscany. Passing through South and Mid Germany, Southern and Eastern Russia, and South Siberia, it penetrates to the Orient shores of Asia, at Ning-po, China. It is likewise abundant in Algeria, and is found all along the northern coast of Africa.

CHAPTER III.

WALKING-STICKS AND WALKING-LEAVES (PHASMIDÆ).

General Peculiarities.

THESE insects, constituting the family Phasmidæ, are amongst the most curious of natural objects. They are most extraordinary in appearance; even more grotesque than the Mantidæ. Frequently they are of great size, some attaining nine inches, and a foot in length: their variety of form is almost infinite. Their names, both popular and systematic, arise from their singular resemblance to vegetable structures; some, long and cylindrical, look exactly like sticks or stems of grass; some might be mistaken for moss-grown twigs; some for lichen-covered bark; while others are invested with spines, like thorns. The imitative resemblance of those known as the Phylliides to leaves is marvellous. It will be well first to take a rapid survey of the more prominent features of their remarkable body.

Contrary to what exists among the Mantidæ and other carnivorous Orthoptera, the head is ovoid, thick, and has a horizontal direction, or depressed towards the base, the mouth always directed somewhat forward; with the eyes more or less prominent, but ocelli in the majority of species wanting. While the prothorax is always small, shorter than the head, undergoing, strange to say, but little elongation even in species most linear and elongate in form, the mesothorax often assumes an extraordinary length. It may be six times as long as the prothorax, and generally likewise exceeds the length of the metathorax. This extension relatively to the other two thoracic segments is peculiar, since in other groups where it occurs there are powerful mesothoracic wings, whereas the Phasmidæ are noted for the absence or curtailment of these particular appendages. When present, this segment carries them and the second pair of legs only at its posterior extremity, and in like fashion the third pair of legs is attached to the metathorax. Except in the case of the Phylliides, the hind body, or abdomen, is also elongated. Entirely ambulatory, the legs vary much in the details of their shape. Ordinarily the anterior pair is the longest, and the femora often have the basal part compressed; and they are so formed as to stretch out in entire juxtaposition in front of the head, concealing it in large measure, and entirely enclosing the antennæ. There is an arolium or membranous cushion between the claws of the five-jointed tarsi, enabling them to adhere firmly to plants.

We have seen that the tegmina, or elytra, as the fore wings are called, are usually of small size or absent, and when present are attached at the posterior part of the mesothorax. Most often they have the form of scales, and cover merely the base of the wings, and are coriaceous, generally raised in the form of a tubercle, and opaque. The lower, or true wings, on the other hand, may be very large, extending in repose, in a limited number of species however, as far as the extremity of the abdomen, without exceeding it, and are attached to the anterior part of the metathorax. This attainment of a greater development of the wings than the elytra is opposed to what obtains in the other families. When one of the pairs of appendages are wanting, it is on the elytra the complete atrophy falls. In the genus Aschipasma, for example, one finds the wings fully developed, but elytra none; and in Phantasis we discover vestiges of hind wings, but of elytra similarly no trace whatever. True, an exception occurs among Phyllium females, where the upper wings become developed and the lower ones dwindle away; but this is for a special purpose, which will be dealt with later on.

In the absence of wing-covers of a size adequate to protect the normal wings in repose, it is essential that provision be made for their defence. This is effected by the outer margin of the hind wing itself, which is greatly thickened, serving as a flat plate or sheath to the greater part of the wing. Beneath it the wing is folded longitudinally in a complicated fan-like manner, and reposes on the back; and seeing the narrow condition of the body, this coriaceous sheath-like portion must be narrow too, in order to adapt itself to the surface of the abdomen. It looks as if it were really a tegmen; moreover, this appearance is enhanced by the fact, that it is often quite differently coloured from the rest of the organ. In some species it is green, like the short wing-covers, whilst the other part of the wing is pink. Among many, however, both organs of flight exist in a more or less rudimentary state (see Fig. 17), and many more remain throughout their lives without ever acquiring wings or wing-covers.

The antennæ are very variable; and the same remark applies to the colour of the body in many Phasmidæ, which may change from brown in early life, to green, subsequently returning to the brown tint. If this be owing to the presence of chlorophyll or other plant-juices among the insect-tissues, its explanation is not far to seek.

Characteristics and Habits.

Chiefly inhabitants of tropical countries, these insects are extremely sensitive to cold; the occurrence of frost puts an end to their existence. They live on vegetation, lying close to leaves and the branches of low shrubs, and are strictly herbivorous. They devour the leaves, and especially the young glutinous or gummy shoots of the plants on which they reside, and with a voracity so excessive that a single pair will destroy a great quantity of foliage, so that in some parts of the world where they abound they become very injurious. This occurs in the South Sea Isles, in the case of *Graeffea coccophagus*, a brown slender species, which sometimes commits dreadful devastation in the plantations of cocoanut trees, occasioning scarcity of food, and orders have been issued by the chiefs for their destruction. One writer goes so far as to ascribe the cannibalism in some of these islands to the want of food caused by the ravages of this insect. *Diapheromera femorata*, common over the greater part of the United States, has also on several occasions appeared in such numbers as to be seriously destructive to foliage in the forests. But taken as a whole, they are far from abundant enough to do any real harm.

Immobility; the Reason and the Use thereof.

Their large size notwithstanding, they are timid inoffensive creatures, and of sluggish mode of life, as their structure indicates. Their body is much too linear, and too long, in the majority of species, to be adapted for rapid walking; indeed, the entire organization of the greater number is such as to impede, rather than to give play to, their means of locomotion. The legs themselves, feeble and of inordinate length, sometimes almost thread-like and very fragile, form obviously far from perfect ambulatory organs. When their owner attempts to stand upright it appears in a state of unstable equilibrium, and has a curious lateral swaying motion, much like a rope-walker, that is most ludicrous. Among the slenderest species the limbs seem to fulfil better the functions of grips or catches than of ambulatory organs, permitting of the insect getting from one branch, or from one bush to another, by enabling it to lay hold of distant supports. Motion for the apterous species, then, is slow and laborious, and they pass the greater portion of their time in a state of immobility, applied against foliage, only shifting their quarters through exigency of obtaining food. Doubtless there are forms among them belonging to the type described, and to other types, that approach more nearly to what we may call the normal form, whose bodies are less slender, their legs shorter and less attenuated, and better fitted for walking, but none the less they are sluggish insects, which live clinging to the branches and leaves of plants.

Winged species (see Fig. 12), like the giant Acrophylla of Australia, which has a comparatively stout body, and the different kinds of Tropidoderus and Podacanthus, bulkier still, and shorter, have greater facility of move-

FIG. 12.—*Tropidoderus rhodomus*.

ment; but the large wings are employed rather momentarily than for anything properly called flight, merely as assistance in leaping to a safer place, rather than for transport to long distances.

Defenceless, and incapable for the most part of rapid movement or of flight, the Walking-Sticks have no other means of escape from their enemies than to pass undetected, in concealing their existence from other animals by their immobility. It is this immobility that renders effective the extraordinary and important characteristic that distinguishes them—their imitative resemblance to vegetative objects. Thanks to it, they succeed in effectually deceiving the eye of the sharpest enemy. Little as it might appear, they have considerably less to fear birds as foes, by which some are much relished for food, than small animals that prowl among the bushes, particularly lizards; and among insects the Mantidæ, which live in the same spots, and are armed for battle in a manner that no Phasma could resist. Several bugs, too, suck the Phasmidæ. One species is known to have harboured Ichneumon-flies in its body without suffering any apparent harm from their presence or their emergence. There is perhaps no other group of insects which is so generally imitative, and which naturalists have experienced greater difficulty in detecting in their haunts, a difficulty heightened by their habit of living solitary, or in pairs. Apterous individuals, such as Diapheromera, however, seem to possess gregarious tastes, often a whole colony being found clustered together, distributed over the branches of the same bush. Having succeeded in discovering one, the investigator, to his surprise, is not long in distinguishing others, which, owing to this imitative resemblance, he may have seen for a long time, have even examined with care, without recognizing anything out of the way, mistaking the living insects for the dead branches.

More Means of Defence.

Putting on one side for the present this wholesale Mimicry of the Phasmidæ, with its protective value, we find some, at least, of the insects in the possession of peculiarities which perhaps one ought to consider as means of defence. The prickles and spines with which the Heteropteryx, Extatosoma, and other ugly monsters are clothed must make them somewhat formidable morsels for insect-eaters to assail; and there are many which have even a more potent means of defence in the power, to a greater or less extent, of ejecting a nauseous fluid. One species is named *Phasma putidum*, from the offensive nature of the secretion discharged. A sluggish creature common in some of the United States, Anisomorpha will, if seized, emit a vapour that slightly burns one's skin; and the highly acrid fluid squirted out by *Graeffea coccophagus* when alarmed, causes great pain, and sometimes blindness, when it strikes the eyes: one of the South African species is said to be able to eject its fetid fluid to a distance of five feet. This liquid comes from two glands or pores placed in the prothorax, more or less apparent according to the species. In some, as among Anisomorpha, they are of exceptional size, quite occupying the sides of the thorax.

For some of the Phasmidæ there appears to be another safeguard—for those remarkable species which constitute the rare examples which are believed to possess aquatic habits. An odd Brazilian species, one of the Prisopi, has the peculiar habit of hiding under stones submerged in the mountain streams, being enabled so to do by the hollowed lower side of the body, and the dense fringe of hairs with which it is in various parts beset; it is supposed to expel the air from the body so as to adhere to the upper surface of a stone. A still more curious insect, probably allied to the same genus, and found in the Isle of Taviuni, seems even more profoundly modified for an aquatic life. Along the lower margins of the sides of the metathorax there stand straight out five conspicuous fringed plates, said to be a kind of branchiæ, or tracheal gills; these coexist with tracheæ opening by stigmata on the exterior of the body for aërial respiration.

Curious Power of reproducing Lost Limbs.

A curious and interesting thing about these creatures is their power of reproducing lost or injured limbs. Limbs fragile and so long look liable to be broken, especially in the insect's early stages of existence. That the accident frequently happens may be judged by the numbers of individuals one sees in collections having one of the legs disproportionately small, though perfectly formed throughout; and sometimes a specimen is met with with the two corresponding legs thus abbreviated. Experiment shows that if during growth—that is, at any time previous to the final moult—a leg be mutilated beyond the base of the thigh, the whole leg as far as the base of the thigh is dropped before the next moult, and at this moult is renewed, either as a straight short stump, in which the articulations are already observable, or as a miniature leg; in the former case, the leg assumes at the next moult the second aspect; this latter form being always changed at the succeeding moult into a limb practically normal in every respect save in this—it never gains its normal size. If the injury to the leg be nearer to the body than the base of the thigh, no reproduction is effected.

Yet another interesting feature of the walking-sticks should be mentioned—the frequent extreme difference of the sexes. There is a general amount of resemblance between the males in being usually smaller, slenderer, and furnished with longer legs and antennæ; the females being generally more robust and bulky, and with shorter limbs. Strange to say, the former often possess full-sized wings, while they are quite wanting in the other sex. When there is a difference between them as to the organs of flight, they are more fully developed in the male. On the other hand, the resemblance to portions of plants is greatest in the female sex.

Remarkable Nature of the Eggs.

But of all the strange characteristics of the Phasmidæ, none is more strange than the eggs they lay. Certainly they are of a most remarkable nature, and very different from insects' eggs in general (see Fig. 13). Hardly any one sees them without observing their extreme resemblance to seeds. It has been suggested that this is for the purpose of deceiving Ichneumons; but if it be so, the imposition may fail of effect, since the eggs are known to be actually destroyed by Ichneumons. Those of *Diapheromera femorata* are flattened and elliptical, resembling beans, with an oblique yellow punctured lid or cap at one end; they are brown, with one side—which shows an elongated pit—banded with yellow. In *Bacteria cornuta* they greatly resemble leguminous seeds; the little

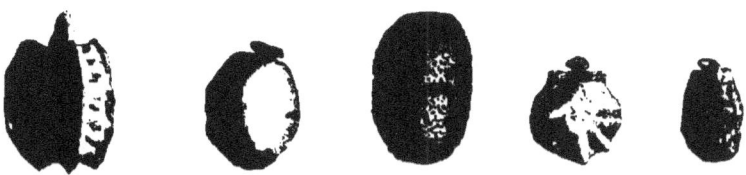

FIG. 13.—Eggs of different walking-sticks.

operculum at the end is distinct on a smooth edge which it exactly fits, the rest of the surface being variously impressed. In some species the cap becomes a sort of knob, and carvings of different designs are seen on the sides of the egg itself, particularly about the part that presents the sunken pit. These eggs are deposited in no careful way, not secured to any object, but, it is worthy of remark, are shed like seeds, being dropped at random loosely on the ground; for the mother, unlike most orthopterous insects, makes no provision for their safety. The noise caused by the dropping of the eggs of *Diapheromera femorata* from the plants on which the insects are feeding to the ground, might be mistaken for the pattering of rain. Thus unprotected, the eggs of this species sometimes lie till the second year before hatching.

Yet in a certain way the egg is protected, for each egg is really a capsule or sac containing an egg, and the lid or cap with which it is provided, already referred to, is pushed off by the embryo when about to hatch.

Although the present state of investigation does not admit of speaking definitely on the point, the generally received opinion is that the egg-capsules are formed in the egg-tubes, only one egg being formed at a time in a tube. This capsule has induced some investigators to believe that the egg contained therein is really a pupa; that all the larval stages are undergone in the egg, and that the insect after emergence should be looked upon as an active pupa that takes food. Comparatively few eggs are produced, from twelve to twenty or thirty being the usual number, but more have been noticed in the case of Diapheromera. Laying takes place in the autumn, emergence between the months of May and August.

Perhaps of all the eggs, those of the walking leaves are the most curious, and their resemblance to seeds is especially striking. At different times specimens of Phyllium have been introduced into Europe, and raised from the egg, affording entomologists the opportunity of studying them. The egg-capsules of *Phyllium scythe* are of the size of a fair-sized pea, though not of that shape. "If," says Murray, "the edges of the seed of *Mirabilis jalapa* were rubbed off, the seed might be mistaken for the egg." What may be called the back and sides are deeply ridged, and the front is flat, with a slender fusiform plate on the middle; in other words, all the ribs are about equi-distant, except two, which are wider apart, and the space between them flatter, so that when the egg falls it rolls over until it comes to this side, and so lies. All except the front is pierced, as it were, with holes, giving the porous aspect of the bark of trees. At the top a tiny conical lid, roughly resembling a Phrygian cap, fits tightly to the mouth. On removing the lid a beautiful white chamber is espied, smooth like porcelain.

The resemblance to seeds displayed by these eggs extends not alone to appearance and to shedding, but even the minute structure of the capsule bears a close resemblance to vegetable tissue. It has been examined by several entomologists; and Henneguy, who enters into some detail in his account of the eggs of *Phyllium crurifolium*, says, "Almost every botanist, on examining for the first time a section of this capsule, would declare that he is looking at a vegetable preparation."

The Scramble out of the Egg; and After.

We will suppose the young phasma in the egg to have acquired its six legs and to be ready to hatch. In the egg it is packed away in a truly marvellous manner. In the full-grown insect, we know, the three parts of the thorax, each of which carries one pair of legs, are of very unequal length, the first pair of legs being borne by the wonderfully short prothorax, as compared with which the meso- and meta-thorax are remarkably elongated. But in the egg this great difference of length of the three divisions does not exist, so that the legs are not very far apart, and pack away closely. But the instant the creature has fairly escaped from its prison-house, the egg, the usual difference in the length of the several thoracic segments is attained ; much expansion of the body comes about during the withdrawal out of the egg, so that it is hard to understand how it was contained therein ; it looks, indeed, like a juggler's trick.

Of the subsequent development of the insects, practically we know very little. But the observations made indicate great differences in the length of time occupied by it, and in the number of moults. Some species are stated to moult many times ; *Diapheromera femorata* is said to reach the perfect stage in six weeks, and to moult only twice ; while according to Murray, who had some specimens under his observation during the whole course of their development, Phyllium takes no less than fifteen or sixteen months to complete its growth, and during that period undergoes only three moults, the first of which does not happen until the tenth month after hatching ; it is interesting to learn that the first moult observed took place when the young Phyllium had attained the length of an inch. It may be mentioned that nothing is more difficult than to watch all the successive moults of an insect of the orthopterous order, since it is their habit—and the habit has been remarked in Phyllium—to devour their skin almost as soon as it is shed; all trace of the occurrence will be over in the space, say, of half an hour.

The day previous to the transformation the young Phyllium showed signs of great agitation, and the body was subject to repeated shakings, eventually ending in the rupture of the skin. At each change of skin there is an immediate increase in size, similar to the enlargement occurring on emergence from the egg, each limb becoming about a fourth larger and longer than the corresponding portion of the envelope out of which it has that moment been drawn. In Phyllium the abdomen especially enlarges after each moult. When freshly hatched it is of a reddish yellow, like a half-dried leaf; for though the colour varies at different periods of its life, it always more or less resembles a leaf. After it has settled to eat the leaves it speedily becomes a beautiful bright green. This colour, as the season advances, gets mixed with yellow, almost passing to the tint feuille-mort, suggesting autumnal foliage, or at least a decaying leaf, agreeable to the very tints which the leaves go through themselves.

In no group of insects as a whole is it more difficult to distinguish the young and the adult states. Among the winged species it is not always possible to distinguish on a young form whether it will, or will not, in course of time, receive the organs of flight. According to Murray, the wings in Phyllium disclose themselves in the period of youth by slight swellings on the meso- and meta-thorax, but it is very doubtful if this character is always well indicated. In Phyllium the organs remain of very small size till the third moult, which suddenly liberates them in their full development; they are drawn out of little cases about a quarter of an inch long, and a few brief moments suffice for them to attain their perfect size of about two and a half inches; half an hour after the last moult the insect is fit for flight. The same may be said of their long antennæ, which are twenty-four jointed as possessed by the adult males, contrary to the females, which have them much shorter, consisting of but nine joints. But these organs remain rudimentary in the two sexes during the young period: it is only in the last moult that they acquire among males their normal length, suddenly shooting out with twenty-four joints. Among the subapterous insects there are species which keep in the adult state the wings in the immature form, analogous to that of the young insects. We cannot, therefore, from the appearance the rudimentary wings assume deduce any positive character, that can help us to distinguish, in every case, the larvæ from the adults. We can only affirm that, as often as the elytrum or the wing is separate and articulated, the individual is an adult. In a great number of the apterous species it is often quite impossible to distinguish the larvæ and nymphs from the perfect insects.

CHAPTER IV.

WALKING-STICKS AND WALKING-LEAVES (PHASMIDÆ)— *continued.*

FIG. 14.—*Acrophylla titan.*

Marvellous Imitative Resemblance.

WE come now to speak of that wonderful characteristic of the Phasmidæ, which renders them amongst the most singular of known insects, which is most likely to attract the stranger, from which they have been given their names—their imitative resemblance to vegetative objects.

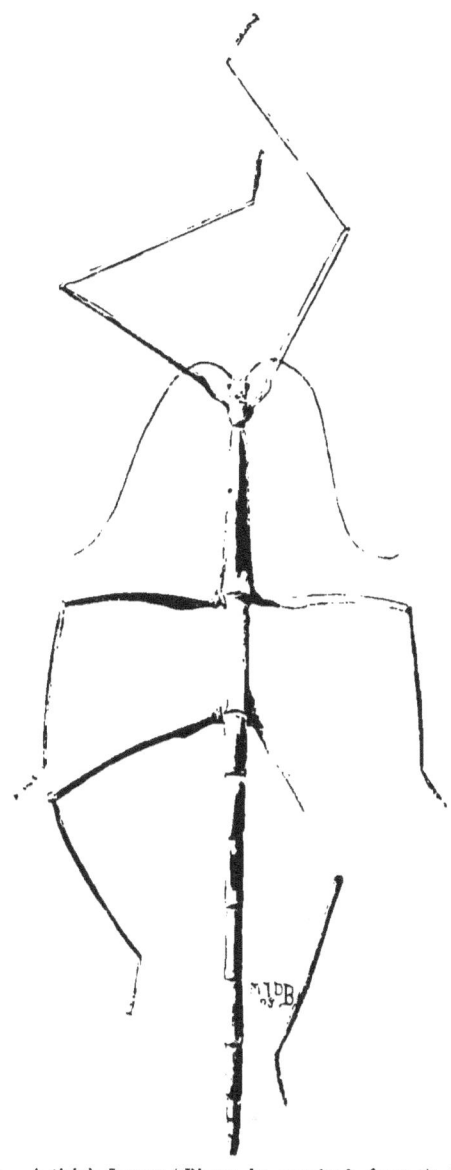

FIG 15.—A Stick Insect (*Phanocles curvipes*), from St. Vincent.

The perfection of this resemblance in certain cases one cannot conceive; it is marvellous; no insects display this kind of imitation so perfectly. Those with slender body, long and cylindrical, so as to resemble sticks, might be mistaken for the latter with all the minutiæ of knots and branches, formed by the insects' legs, which may be stuck out rigidly and unsymmetrically. Dr. Wallace, the naturalist, familiar with them in tropical forests, describes them in the Moluccas " hanging on shrubs that line the forest paths; and they resemble sticks so exactly in colour, in the small rugosities of the bark, in the knots and small branches imitated by the joints of the legs, which are either pressed close to the body or stuck out at random, that it is absolutely impossible by the eye alone to distinguish the real dead twigs which fall down from the trees overhead from the living insects." And he adds that he has " often looked at them in doubt, and has been obliged to use the sense of touch to determine the point." Some are small and slender, like the daintiest of straws or twigs; their body sometimes an inch and a half, sometimes barely an inch long, the legs like threads; others are of a much larger and stouter kind. The larger wingless sticks (see Fig. 15) are often eight inches to a foot long. Many of these are hardly thicker than a knitting-needle. In Mexico, for instance, Phanocles is about eleven inches long in all, and exhibits the odd knitting-needle effect, enlarging nowhere in circumference except at the base of the legs. Some species of Bacteria are so excessively slender that the linear body is scarcely as thick as one of the legs it bears.

Others, again, can fly, having ample wings, and, oddly enough, often gayly coloured. Look at the large spectre *Acrophylla titan* of Australia, a giant of its kind (see Fig. 14): its charming wings generally blackish brown in colour, but irregularly spotted and banded with white, the costal portion variegated with green and pink, and expand fully eight inches; while the long cylindrical body itself exceeds this length, and is as thick as a man's little finger. Some kinds of Tropidoderus and Podacanthus, also from Australia, likewise expand eight and even nine inches in flight, and have a stouter though shorter cylindrical body. *Podacanthus typhon*, the pink-winged spectre, is one of the showiest species, owing to its size and the fine pink colour which tinges the hyaline wings; while the front portion of the wings, like the short tegmina, is of a grass-green: but when at rest the immense and brilliant wings fold up like a fan, so that the pink colour is completely concealed under the narrow front and wing-covers, and the whole stick-like insect is then green or brown, and almost invisible among the twigs or foliage.

Bizarre Shapes Galore.

Almost all the foregoing forms are smooth-bodied, or have merely little insignificant inequalities of skin. But to increase the resemblance to vegetation, some of these Phasmas have the oddest unequal surfaces imaginable, so that they resemble the roughened bark of the trees among which they live, or bits of rotten wood, or lichen-covered bark. There are bizarre appearances among them which defy description. They have growing to them or sprouting from them almost everywhere, more especially from the abdomen and legs, delicate small green processes or foliaceous excrescences, looking exactly like the Hepaticæ, or moss. They inhabit damp forests both in the Malay Isles and America, and it needs careful scrutiny to detect that the apparent piece of rotten moss-grown twig is in reality a living insect. See *Ceroys laciniatus* (Fig. 16) from Nicaragua; how these irregular leaf-like expansions protrude all over it! Could one wish for anything more grotesque than the bulky, prickly, spiny, briefly-winged giant Heteropteryx, from India, Borneo, Sumatra, and Australia, or than a Peruvian Ceroys, typifying thorny stems? Or, turn to the dilated-bodied spectre Extatosoma, hailing from Australia, from Tasmania, and New Guinea, enormously thick as compared with the males, and with both the wings and the wing-covers rudimental, like tiny sprouts. The body is spinous, furnished with membranous lobes or dilations at the side; it is spined on the thorax, the abdomen is spined, so are the dilations of the legs, and the peculiar small pyramidal head is strongly spinous at the apex.

End gained by this Mimicry.

This armature, these details of form and colouring, may in all cases be regarded as developed for protective purposes. Protection in resemblance, and thus concealment, is a very general and very effectual means of maintaining life in the ceaseless struggle for existence. Walking-sticks being perfectly inoffensive and herbivorous, and therefore needing no special qualification for facilitating capture of prey, their mimicry is purely defensive, its ultimate end to elude their enemies. Obviously, their harmlessness, their solitary instinct, their sluggish motions, as a rule, and incapacity for what can be properly called flight, their soft and succulent nature, so that they are eagerly devoured,— render them particularly open to attack. Their defence, their very existence, depends upon their being by form and colour concealed from enemies. That their vegetable disguises deceive their natural enemies the numbers that escape destruction prove: without them, they must soon be exterminated. There is no family to which this protection is more indispensable, and in none, perhaps, is it so generally and so perfectly possessed. Given a walking-stick hugging the stem of a bush or a leaf, with its two hinder pairs of legs stretched straight alongside the body, the front pair outstretched in the opposite direction, and the antennæ snugly tucked between them, the deception immeasurably increased in efficacy by its immobility, it must be a sharp-sighted creature to ever discover its presence. Even those provided with the most splendid and expansive wings use them rarely, and that use, as we have seen, is of the briefest description. Whenever they settle the great wings close; their beauties are no longer displayed. An insectivorous foe spying such a walking-stick in flight, and making for it, would search in vain for what he had seen; the gorgeous aërial being alighting, had suddenly transformed itself into its more habitual unobservable stick shape.

Leaf-Insects.

All these Phasmidæ are noted for their lank and usually slender bodies and legs. But there is an exceptional form of walking-stick. These have remarkably expanded bodies; they are known as Walking-Leaves, from the striking resemblance they present to the leaves of trees. While seated among the leaves of the trees on which they live, no more exact representation of a growing leaf could be conceived. Not one person in ten can see a leaf-insect when resting on the food plant close beneath their eyes. It is principally the tegmina or front wings that display this great resemblance in Orthoptera, and in none of the order is the extraordinary phenomenon more marked than it is in these Walking-Leaves, of the genus Phyllium. The genus constitutes by itself the tribe Phylliides, the members of which belong exclusively to the tropics of the Old World, coming from the Philippine Isles, Java, and Ceylon; in fact, they extend from Mauritius and the Seychelles even as far east as the Fiji Isles, having, it would appear, a strange penchant for insular life. Although the group has been very inadequately investigated, some twenty species are known, and the

FIG. 16.—*Ceroys laciniatus*, from Nicaragua: irregular leaf-like expansions protrude all over it.

individuals are believed to be not uncommon, notwithstanding the limited number of specimens that our collections contain.

As will be seen from Fig. 18, the resemblance to a leaf of the tegmina is of the most remarkable nature, and the other details must add greatly to the deceptive appearance in their native haunts. The imitation of a leaf is carried out with a degree of exactitude so surprising, that it seems, in truth, that the whole insect has been shaped and charged with appendages to produce the perfect resemblance. The body, which escapes completely from the family type, has become large, oval, and depressed; the antennæ have become extremely short, and are flattened; all the legs compressed and dilated in leaflet form; the hind wings atrophied, being represented merely by a minute process; while to the broad leaf-like tegmina lying flat on the back, the head and small prothorax together form, as it were, the swollen petiole or leaf-stalk, on both sides of which the flat leaf-like expansions on the front legs answer admirably as stipules. The insects feed only at night, resting motionless during the day, aiding them to elude detection. During the early stages, when the insect does not possess the tegmina, it is said then to adapt itself to the appearance of leaves, by the movements it makes, and the positions it assumes augment the resemblance.

In these wonderful Eastern leaf-insects of the genus Phyllium, it is the female only that so marvellously imitates a green leaf; she alone is possessed of the large leaf-like tegmina. The males of the genus are altogether different from the females; having instead of the foliaceous tegmina, short wing-covers that are not leaf-like, while the gauzy hind wings, which are particularly large and conspicuous, are totally devoid of leaf-like appearance.

Tegmen of the Female an Exceptional Structure.

This disguise of the female in respect of its tegmina is striking to a naturalist from various points of view. He will notice that whereas when there is in insects a difference between the organs of flight of the two sexes, the male has them largest, the very opposite is true of Phyllium; that is to say, the normal condition is adhered to so far as the hind wings are concerned, but in the front pair the rule is reversed, the leaf-like tegmina of the female exceeding greatly the rudimentary wing-covers of the male. He will also observe as one of the peculiar traits of the family, a trait shown by all other members of the tribe, that the wing-covers or tegmina (when they exist) are greatly abbreviated, even when the wings are largely developed. This is the case in the male of Phyllium, but the female offers precisely the opposite character — the wing-covers, the only members susceptible of such exact mimicry of a leaf, being greatly developed, while the wings are aborted. There is this to be pointed out, that had the wings, and not the tegmina, been made to resemble a leaf, the mesothorax would have remained entirely visible; it would have spoilt the perfection of the resemblance of the insect and a leaf, by forming with the head and prothorax a petiole, as it were, too large, and too apparent. Nature has then, contrary to the general rule, atrophied the wings and developed the tegmina, to obtain the appearance of a leaf over the largest surface of the body possible. The great reduction of size of the prothorax and antennæ, parts which would equally interfere with the resemblance, is likewise to be included in this artifice, tending to the same end.

Another surprising thing—it is probable that the female tegmen of Phyllium is a structure as peculiar morphologically as it is in other respects. In order the better to imitate a leaf, the radial vein is placed quite close to the posterior edge, permitting the radial veins of the tegmina, when they come together, to juxtapose, typifying the principal nerve or mid-rib of a leaf. The lateral ribs of the leaf are represented exactly by the oblique costal veins. It will thus be plain, that the tegmina of female Phyllium not only break a rule that is almost universal in the Insecta, and reverse the normal condition in the family to which they belong; but also differ widely from the same parts of its mate, and, moreover, are completely different—in quite an exceptional manner for a Phasmid—from its own other pair of organs of flight.

This extreme resemblance of Phyllium to a leaf has attracted the notice even of the natives of the tropics where they abound, where little or nothing is known of natural history. In many such places, as in the Indies, it has given rise to the singular belief that the insects are truly transformed leaves, by which the inhabitants understand a bud developing into a leaf, and subsequently being converted into a walking-leaf insect. We have evidence of the idea obtaining credence in Ceylon, and no explanation could shake the rooted conviction in the reality of this miracle.

When first brought into notice in this country they created unbounded surprise. Thus Richard Bradley, a fellow of the Royal Society, and at one time Professor of Botany in the University of Cambridge, writes of them in 1739, in a philosophical work, in a fashion fully as grotesque as any legend of the countries which they inhabit. This fanciful author regards them as exhibiting identity of animal and vegetable development, "being nourish'd," he observes, "as well by the Juices of the Tree, which the Mother Insect lays its eggs in, as by its own." This bald statement he gravely gives in more elaborate guise: "The Insect, when it has found its proper Tree of Nourishment, lays its eggs separately in the Buds of it, which hatch when the Buds begin to shoot; the Insect then is nourished by the Juices of the Tree, and grows together with the Leaves till all its Body is perfected; and at the Fall of the Leaf, drops from the Tree with the Leaves growing to its Body like Wings, and then walks about; this is not common enough with us to be easily believed, and what I should not have ventured to mention in this place, if the Insects themselves were not to be met with in the curious Cabinets of our own Country.

"What I account the most curious point belonging to this Relation is, That the Sap of any Tree should be so naturally adapted to maintain at once both Vegetable and Animal Life; and by that means to unite the Parts of two Beings, so distinct from one another as Plants and Animals, and circulate the same Juices equally in the Vessels of both. . . . That a Leaf of a Plant should so unite itself with an Insect as to make one distinct living Body is wonderful."

All one can say is, the remarkable appearance of the insects affords some excuse for the absurdities of this romantic story.

Somewhat more than a century later, the eggs of one of these insects were introduced from India to Edinburgh, where they hatched, and the living insects were under the observation of the late Mr. Andrew Murray for nearly eighteen months in the Royal Botanic Garden. Strangers used to search in vain to distinguish Phyllium from the surrounding leaves. Mr. Murray says of one: "For the greatest period of its life it so exactly resembled the leaf on which it fed, that when visitors were shown it they usually, after looking carefully over the plant for a minute or two, declared that they could see no insect. It had then to be more minutely pointed out to them; and although seeing is notoriously said to be believing, it looked so absolutely the same as the leaves among which it rested that this test rarely satisfied them, and nothing would convince them that there was a real live insect there but the test of touch. It had to be stirred up to make it move."

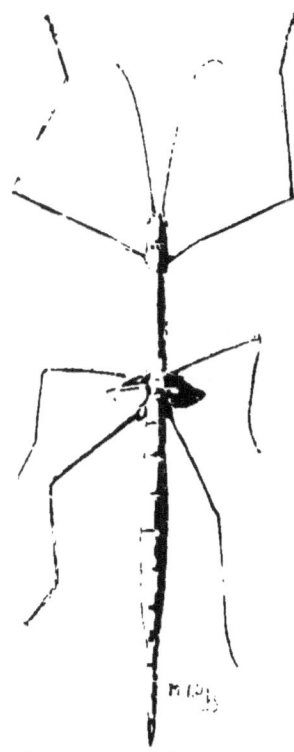

FIG. 17.—*Graeffia cocephagus.*

The same was true of some specimens exhibited alive

in 1867, at the Jardin d'Acclimatation at Paris. They were placed on a growing bush, from which most of the leaves were stripped, that they might be more easily perceived. If a large placard had not announced the presence of the insects, no one would have recognized anything extraordinary; and as it was, many persons, after carefully examining the case, went away without being aware of having seen anything but the bush, convinced that the placard referred to something microscopic, too minute for their sight. Even those who were acquainted with the aspect of walking-leaves had often a long search to discover what was in reality before their eyes.

FIG. 18.—A Walking-Leaf (*Phyllium scythe*), from Silhet.

Distribution.

Not the genus Phyllium alone is exotic, the whole family of Phasmidæ is eminently tropical. About six hundred species of the family are known—by the way, a number small in comparison with that in many of the large families of Insecta—but only four or five are found in Europe, slender apterous green or brown kinds, measuring but two or three inches in length; and they are all restricted to the south, only one reaching as far north as Central France. They belong to the genus Bacillus, and while common enough in the female form, in comparison the males are extremely rare. In the temperate regions of America we find similar restriction.

In the warm portions of the globe Phasmidæ are of almost universal distribution. Formerly India and the isles of the Indian Archipelago were regarded as constituting the metropolis of the group, but recent researches seem rather to point to Australia as the region where they are now most largely developed. A species of Podacanthus is so common, that it is rare in the summer-time in any part of Australia to find a gum-tree without a few feeding upon it; and occasionally the trees for miles will be denuded of their foliage by it. New Zealand has several species of Phasmidæ, differing from the Australian forms, and all wingless. Of America it may be remarked, that not only is it less rich in Phasmidæ than the Oriental hemisphere, but the species there have a tendency more pronounced to remain apterous. Among what may be called wingless species of the Orient, more usually the rudiments of wings are apparent, or the wings are more or less developed, although of the briefest description. From these differences, and others, it follows that the genera of the different continents are in great part distinct. Not only is this the case, but the greater number of species are confined within somewhat narrow geographical limits. It is easy to understand how insects so little addicted even to walking do not spread rapidly, and that, seeing their absolute dependence on plants which serve at once as their abode and their nutriment, they are affected more strongly than others by the modifying influence of the places which they inhabit; whence it comes that each little region possesses its species.

Existence is possible to the Phasmidæ only in verdant regions, where delicate nourishment is always to be had within reach. They could not live in a dry climate, and on a soil often deprived of vegetation, their limited powers of locomotion being an insurmountable obstacle to their travelling over great distances in search of food. These facts explain at once the rarity of phasmidæ in tropical Africa, and the manner in which they preeminently affect the humid and verdant isles of the tropics, and it is for those of the Indian Archipelago, as before mentioned, they show partiality. Even species with large individuals seem to be able to continue their existence in comparatively small isles.

That Walking-Sticks come of a Remote Antiquity.

Few groups of insects are so specialized; and one would naturally expect these bizarre creatures to be the outcome of a long series of forms in a special line of development. Some insects, said to belong to the genera Phasma and Bacteria, have been found fossil in one or two fragments in amber, belonging to the early part of the tertiary period. No phasmid has been unearthed from the great formations of the mesozoic period, so that, with the exception of a single insect-fossil from the tertiary strata in North America which has been recently referred to this family, but probably in mistake, we know nothing of fossil walking-sticks older than the remains preserved in amber. It must be stated, however, that the upper coal-measures of Commentry, France, have suddenly revealed a considerable number of forms that may be connected with our living Phasmidæ, a discovery carrying these Orthoptera back at once to the remote antiquity of paleozoic times. M. Brongniart describes two remarkable gigantic insects from the carboniferous beds. Both he and Scudder have treated these fossils as forming a distinct family called Protophasmidæ. So far as we know them, these early types differed from those of to-day by being invariably winged, and in that both pairs of wings were adapted for flight; the front pair—what we now call wing-covers, or tegmina—being not leathery and thickened, as now, serving as mere protective flaps to the closed hinder pair, but were as large and diaphanous as their posterior fellows: these ancient insects explain the origin of our living giants, being twenty-five to fifty centimetres long, and as much as seventy in spread of wing. To them have been referred, on inferential grounds, a whole group of detached wings found in carboniferous beds in Europe and America. They further differed from modern types in having the several parts of the thorax more nearly equal in length—similar, in fact, to the condition while still in the egg of walking-sticks of to-day; thus illustrating once more, what many naturalists believe, that in the development of the individual we may trace, more or less completely, the ancestral development of the race.

The Phasmidæ are a singularly isolated group; we discover no transition properly so called between this family and others. So far as appearance goes, they approach most closely the mantidæ; their forms are sometimes very much alike; in several mantids

(Thespis, etc.) the body is even as slender and cylindrical as among Phasmidæ. Yet there is no real affinity between them, essential differences distinguish these groups; differences relating to the modifications of nearly every part of the body, in connection with habits diametrically opposed. Still, one can hardly but say there is a sort of parallelism between them; the two families constituting, so to speak, two collateral series, the one representing the herbivorous, the other the carnivorous type of the same form.

SALTATORIA.

CHAPTER V.

LOCUSTS AND GRASSHOPPERS (ACRIDIIDÆ).

FIG. 19.—A Grasshopper (*Truxalis Pharaonis*).

THE reader will perceive, on referring to the Contents, that the families of the Orthoptera are divided into two sections. He will observe that the Mantidæ and Phasmidæ, already considered, belong to the Cursoria, or Runners, their hind legs being but little different from the others. We come now to speak of two families pertaining to the Saltatorial group, each remarkable for their saltatory powers, due to the great development of the hind legs, which are more elongate than the others.

and have thick and powerful thighs. As compared with the Mantidæ and Phasmidæ, which are amongst the most distinct of any subdivisions of the Insecta, these families are much more intimately allied. Technically, they are known as Acridiidæ and Locustidæ; but, unfortunately, there is a certain amount of confusion between the scientific and common names, the latter not corresponding in their application with the family limits. The names Grasshopper and Locust are almost synonymous. Not only in a popular sense, but even by scientists, the word "grasshopper" is applied at one time to the true locusts or to the various species that constitute the Acridiidæ, and at another to species belonging to the Locustidæ; "grasshopper," in fact, is a collective term used to include most of the species of two different families. Notwithstanding this difficulty in applying the popular name, the distinctions between the insects composing the families are readily recognized. Briefly stated, while the Locustidæ are usually found on the grass, bushes, and trees, and have very long thread-like antennæ, generally longer than the body, and the tarsi four-jointed; Acridiidæ includes those species which generally reside on the ground, they have antennæ of less than thirty joints, never exceeding the body in length, and the feet or tarsi short, with three distinct joints. With the Acridiidæ we commence our consideration of the Saltatoria.

LOCUSTS AND GRASSHOPPERS (ACRIDIIDÆ).

This family contains the majority of the species, a very large number, varying considerably in form and character; it is the most numerous in species and individuals of any of the families of Orthoptera. The small grasshoppers which are common in our fields give a very good idea of their general appearance. Active little insects having a body laterally somewhat compressed, a large head, conspicuous eyes, the hind legs formed for leaping, the wings in repose deflexed and pressed to the sides, our native grasshoppers, however insignificant and unimportant, represent the family quite as truly as do its many more imposing, remarkable, and interesting species.

Anatomy.

In dealing with a family so active as a rule, so famous for the extraordinary movements of some of its members, one turns naturally to investigate their instincts and general intelligence. But it is hard for us to appreciate the intelligence of insects. It depends in them, of course, largely on the development of the organs of special sense; and the study of sensations, if one of the most fascinating, is at the same time one of the most difficult of the departments of entomology. Recent research has resulted in much definite knowledge of the structure of the sense-organs, but there is a great lack of experimental basis for conclusions as to the functions of the various organs described. However, there is a great variation in the degree of perfection of the different senses in different insects.

The sense of sight must be well developed in the Acridiidæ, for they are furnished with two large well-developed compound eyes, and three simple ones (ocelli), supplied with nerves of special sense. The compound eyes are situate at the sides of the head ; in front are the ocelli, one on each side between the eye and the insertion of the antenna ; the third ocellus being in the middle, immediately in front of the base of the antennæ. This family is one of the large groups of insects in which the coexistence of the two kinds of eyes is most constant. The ocelli, however, vary much in their development, being in some cases prominent and easily perceived, while in others they are very imperfect, difficult to detect, apparently functionally useless. In what way the insects specially need two sorts of eyes, is not clear. Later on we shall see that a similar condition in regard to sensitiveness to sound is believed to exist in this family. The antennæ are organs of touch ; and the palpi not only serve the same function, but probably are endowed with the sense of taste.

It need hardly be said an insect breathes by means of a complicated system of air-tubes or tracheæ, ramifying through every part of its structure, and communicating with the air by a row of spiracles, or air-holes, or breathing-holes (stigmata), in the sides of the body. Acridiidæ are remarkable amongst the Orthoptera for the possession of air-sacs or vesicular dilatations in the interior of the body in connection with the tracheæ. Many winged insects possess such vesicles. In bees, wasps, moths and butterflies, flies, and in some beetles, and some bugs, they are found well developed, being most numerous and capacious in volant insects which sustain the longest and most powerful flight. But in the larval or immature forms of these they do not exist, nor do they occur in truly apterous insects. In the flying locusts they are as numerous and as large as in any group, more numerous perhaps than in the bee in proportion to the body, while certainly there is a greater number of large sacs, and both sexes are equally well provided. Packard describes the distribution of these remarkable elastic sacs in the Rocky Mountain Locust. There is a thoracic set, consisting of a pair of very large size, with which are connected some smaller ones in the head; and an abdominal set. The last are very extraordinary, being of such large size that they may be regarded as only becoming fully expanded at a time when the body being comparatively empty, food being wanting, there is no normal distention of the alimentary canal, and the other contents of the body are as yet undeveloped. Although according to one observer they are present even in apterous forms of Acridiidæ, distinct vesicles are absent in the neighbouring groups of Orthoptera. Those Orthoptera which do not take long flights have no need of air-vessels.

No doubt they are connected with the power of flight ; doubtless they assist the insect in its aërial movements. The body of a large grasshopper or locust is naturally of considerable weight, and it seems certain true flight can only be effected when the sacs are dilated and filled with air. That by filling and partially emptying them during the process of breathing the insect is enabled to enlarge its bulk and alter its specific gravity at pleasure, so as to render itself capable of rising and supporting itself on the wing with little muscular effort, was first assumed by Sir John Hunter; and inference and observation confirm the view that the use of the vesicles is to lighten the body, to float up the insect in the air. The precise mode in which they are dilated is not understood. With its sacs in full play, plainly a locust becomes an aëronaut, a sort of balloon.

To this fact are largely due the enormous powers of flight possessed by these insects ; the intimate association of the complex arrangement of air-tubes and air-sacs with the powers of flight is very evident. It will be seen that, once having risen from the ground, the insect can sail for hours in the air, constantly filling and re-filling its internal buoys or balloons, and thus be wafted straight on its course for miles by favouring winds. Meanwhile, scarcely as much muscular force is spent through the day as is exerted during a few vigorous hops. Towards evening, and in damp and cloudy weather, the powers of flight are lessened, owing to the diminished power of respiration. In the possession and use of these air-sacs, locusts may be compared with birds.

When in addition to the sacs we find many expanded or dilatable tracheæ, chiefly in parts of the body where there is not room for the sacs themselves, we can duly estimate the wondrous powers of the locust as an aëronaut.

Intellectually the Acridiidæ appear to be the equals of most other insects, while many are inferiorly endowed. Those that excel them in this respect are the ants and bees or wasps, the social hymenoptera, which have a brain constructed on a higher, more complicated, plan than in the other winged insects. It must be understood that the word "brain" so applied is a mere courtesy term, as the brain of our insects does not correspond to the brain of a vertebrate animal. It consists of a double ganglion, placed in the upper part of the head, the first and largest of the two ganglia therein. Moreover, it is a much more complicated organ than any of the others, having parts which are wanting in them, hence it is *par excellence* nearer to our idea of a brain than any of the other ganglia. The other ganglia are eight, three thoracic and five abdominal. This series of nerve-centres, connected by nervous cords, together constitute the nervous system.

The Gift of Song.

If the Acridiidæ are remarkable amongst the Orthoptera for their air-sacs, they are no less interesting on account of another wondrous possession, which they share in common with the other Saltatoria—their gift of song. Singers in the true sense they are not; they should rather be called minstrels; for, strictly speaking, owing to their peculiar mode of breathing, they have nothing that corresponds to our voice, and it may be accepted, as Aristotle expressed long ago, "that no living creature hath any voice but such only as are furnished with lungs and windpipes." While less famed as instrumentalists than the neighbouring Saltatoria, they are not without their resources in musical powers, and there is evidence that they are of great importance to the creatures, though it is hard to state definitely in what way. In some of the aberrant forms of Acridiidæ, some parts of the structures of the body are clearly subservient to the musical organs, and have every appearance of being specially directed to securing their efficiency. The situation and structure of these sound-producing instruments we will pass over for the present.

The stridulation, or "song," is mainly accomplished by rubbing together the inner surface of the hind legs and the outer surface of the tegmina or wing-covers. These latter parts have considerably elevated or projecting veins, one of which is slightly more thickened, and has a sharp rasp-like edge; the inner face of the hind thighs or femora carries a series of small bead-like prominences; by scraping these upon the rasp-like surface of the veins of the wings the wing is thrown into a state of vibration, and a musical, monotonous, nearly uniform sound is produced. A Stenobothrus, when about to stridulate, plants itself in a nearly horizontal position, with the head a little elevated; the shanks of the hind legs he is apt to draw close within a groove beneath the thigh, evidently made to receive it. The legs are now raised and lowered with a more or less regular and continued motion, the thighs being grated against the firm edge of the tegmina. The wings are the responsive instrument, the viol, to which the leg performs the office of bow in this musical performance. Every movement of the fiddle-bow produces a note, the notes varying in rapidity, number, and duration in different species. Few are aware that every kind of grasshopper has its distinctive note; a practised ear can distinguish the song of even closely allied species.

Scudder, who has given much attention to the subject of Orthopteran music, says in North America "the uniformity with which each species of Stenobothrus plays its own song is quite remarkable." One species, *S. curtipennis*, makes about six notes per second, and continues them for one and a half to two and a half seconds; another, *S. melanopleurus*, produces from nine to twelve notes in about three seconds. In both cases the notes follow each other uniformly, and the movements are less rapid in the shade than in the sun. Scudder has even reduced the notes of several species to a written music.

This stridulation of grasshoppers is specially characteristic of the male. Yet it is not always an attribute with them of the male only. It was for long supposed that the males alone sang, that they alone were endowed with the musical apparatus. Females were indeed perceived rubbing their thighs and wing-covers together, but as they appeared to be destitute of instruments, and as no sound resulted from their efforts, it was concluded that these were merely imitative. It is, however, discovered that musical organs do exist in the females of various species of Stenobothrus. Doubtless they are rudimentary as compared with those of the males, but they are believed to be really phonetic, although the appropriate movement produces no sound perceptible to our ears.

The musical knobs of Acridiidæ would seem to be modified hairs, and Graber mentions the finding in females of the stages intermediate between knob and hair. Much variety exists in the structure of these instruments in different species; in *Stenobothrus lineatus* raised folds replace the musical pegs.

Some grasshoppers stridulate during flight, by the friction of the wings and wing-covers. Whether this has any real importance remains almost unconsidered. It appears to be under the control of the insects, for they often omit it when alarmed. Some species produce a uniform noise during the whole of their undeviating flight; others make it only during the intervals of their course, and seem to stridulate more at will.

Acridian Ears.

If the body of an Acridian insect be carefully examined, there will be found in the majority of the species, on each side of the first abdominal segment, an organ which there is much reason for believing to be of the nature of an ear. It is situated a little over the articulation of the hind leg, close to the spot where the sound is, as above described, produced, and just behind the first abdominal spiracle. These ears vary in form, but consist of a stretched or tense membrane. The membrane may be level with the skin, surrounded by a simple horny ring; or it may be somewhat depressed, a portion of the segment projecting a little over it; and sometimes it is very sunken in the abdomen, the arched and horny sides projecting over it so much that nothing is seen externally save a sort of slit with a cavity or pit beneath it. In the last condition the organ exists in the genera Mecostethus and Stenobothrus, which are among our native grasshoppers. It is usually conceded that this ear consists of a tympanum supplied internally with an auditory nerve and a ganglion, besides muscles, and tracheal apparatus; it is undoubtedly a sense-organ of an extremely delicate nature. It is found in both sexes, as in most of the species of Acridiidæ. The forms in which it is absent are generally at the same time wingless, and destitute of organs for producing special sound.

Yet to decide as to the exact function of these ear-like acoustic organs is a matter of extreme difficulty. We know, from the fact that the insects are easily disturbed, that the sense of hearing must be delicate, and the Acridiidæ with ears are believed to be sensitive to sounds by means other than these organs. This suggests that the purpose of the latter is the perception of special sound. What can this be? Is it the fiddling or stridulating sound which we have seen they produce? Any insect having elaborate sound-producing organs must be supposed to have ears to hear the sound produced by others of its own species. No doubt the effective sound-producing instruments are, apparently, confined to the male, while the ears exist in both sexes. But this need give rise to no difficulty, for it is generally thought that the female has pleasure in the music of the male. The real obstacle at present to the acceptance of these organs as being special structures for the perception of the music of the species, lies in the fact of the existence of the acoustic organs in species that do not,

FIG. 20.— *Teratodes monticollis.*

so far as we know, possess phonetic organs, and are incapable of stridulation in either sex.

The difficulty at once vanishes should these species really produce some sound, though we are ignorant as to their doing so. It is well known that sounds inaudible to some human ears are perfectly audible to others.

This Tyndall has illustrated in his work on Sound. "Crossing the Wengern Alp with a friend," he says, "the grass on each side of the path swarmed with insects, which, to me, rent the air with their shrill chirruping. My friend heard nothing of this, the insect world lying beyond his limit of audition." In other words, as Scudder remarks, sounds become inaudible to many persons when they are derived from vibrations more rapid than 25,000 per second, and when the number becomes 38,000, the limit of human perceptibility is reached. This difference in sensitiveness to vibration of human ears, renders it, of course, more probable that ears so unlike our own as are those of insects may be capable of detecting sounds of a shrillness of which the best human ear can hear nothing. We may perhaps then conclude that these Acridiidæ with ears, and to all appearances dumb, do really produce sounds, though beyond our range of perception, and do so by some method unknown to us. If this be the case, it is probable that the function of these ears is the hearing of particular sounds.

Oviposition; and Philosophy of the Egg-mass.

The details of the process of oviposition of the Acridiidæ are of much interest. The insect has no elongate exserted ovipositor for placing the eggs in suitable positions, but possesses instead four short horny appendages, or gonapophyses, which, from their peculiar structure, are admirably adapted for digging. By means of these, when about to lay, she excavates a hole in the ground; there is no perforation of the soil, the hinder part of the body is merely forced into it. With the valves closed, she inserts the tips into ground, and by a series of muscular efforts, and the alternate opening and shutting of the valves, in this way it is easy to press the earth aside, and in a few minutes nearly the whole of the abdomen is buried.

Along with the eggs a quantity of viscous fluid is discharged, that binds all the eggs in a mass, and when the last egg is laid, the viscous matter continues to be shed, filling up, as with a stopper or cork, the orifice of the cavity. The Rocky Mountain Locust stops the neck of its hole with a compact and cellulose mass of the material, which, though light and easily penetrated, is more or less impervious to water. The operation complete, but little trace is left of it; most often the frothy secretion rises above the burrow of the Algerian migratory locust. When fresh the mass is soft and moist, but soon the fluid deposited dries, acquiring a firm consistency, and forms an excellent protection to the eggs, corresponding to the more definite capsules of cursorial Orthoptera.

The insect in ovipositing prefers a hard and compact

soil to that which is loose, and may be observed at its excavations even on beaten paths. The time required to accomplish this strenuous task will vary according to the season and temperature. In the event of a frost at night, and the insects not rousing from their chilled inactivity till 9 a.m., the Rocky Mountain Locusts have been observed to be scarcely able to make the hole and complete the pod during the four or five warmer hours of the day; but with higher temperature not more than about half that time would be needed.

Their period of ovipositing extends over about sixty-two days, the average number of egg-masses formed being probably three or four, and the total number of eggs deposited about a hundred. The French naturalists recently record a similar fact in Algeria, and have ascertained that *Schistocerca peregrina*, one of the migratory locusts, may deposit eggs at more than one of the spots on which it may alight during its migration.

A careful study of the egg-mass, or egg-pod, of *Caloptenus spretus* and other locusts, will reveal a quadrilinear arrangement of the eggs, not only so as to economize space, but so as to best facilitate the escape of the young. Clearly if, from whatever cause, the upper eggs should hatch later than the lower, or should fail to hatch—as is not unfrequently the case—the exit of the young would be impeded were there no provision against such a possibility. Without touching upon the channel obtained, along which the exit is easily made, it may be mentioned that the posterior or narrow end of the egg of *Caloptenus spretus* points downwards in the egg-mass, so that the exit of the young from the anterior end is thus rendered easier. The quadrilinear arrangement of the eggs is by no means constant, even in the same genus. But even in the pods of those species of Caloptenus which have the eggs irregularly arranged, the head ends commonly point either outward or inward, so that the newly-hatched creature may push out at the sides or through a central space.

Egg-Enemies.

Notwithstanding the mother's care, fearful are the odds against the development of her progeny. Ere yet the locust is born, from the moment it starts life as an egg, enemies stand in wait to cut short its career. The smaller of these animals, belonging principally to its own class, carry on this good work most effectually. We altogether undervalue the usefulness of these tiny foes in helping to keep the locust in check, simply because they are often scarcely perceptible, and their work too often goes on hid and unobserved.

Beetles of the family Cantharidæ hover in the localities where the eggs are laid and insert their eggs in the egg-masses of Acridiidæ, which may thus be entirely destroyed. In North America the so-called Locust Mite proves a bitter enemy to the locust. In the mature form it lives in the ground, feeding upon all sorts of soft animal and decomposing vegetable matter. When the grasshopper fills the holes with its eggs, the mites flourish thereon, creeping into the holes and eating the contents voraciously. A most common and widespread egg-enemy is the Anthomyia egg-parasite. The Muscinæ, owing to the rapidity of their successive generations, destroy large quantities of the eggs of the migratory locust. The flies follow the locusts, and when they settle to lay eggs, they also alight, ready to bore their way to the eggs, with deadly results. It is said the Dipterous genus Idia is incapable of boring into light and sandy soils, and for this reason *Acridium peregrinum* chooses them. Two-winged flies of the family Bombyliidæ are also parasitic in these eggs. Künckel d'Herculais has studied the Bombylid larvæ found in the ova of the devastating locust *Stauronotus maroccanus*. The larva issues from the egg in August, reaches the limit of its growth in October, and passing the winter within the egg-case, is hatched the following summer.

To this work of egg-destruction by insect enemies and parasites, must be added the good offices of various birds, and of some mammals. Hogs are quite fond of locust eggs, and soon learn to search them out.

Of the Escape of the Young from the Egg.

It is usual for all the young in a given mass to burst from the egg very nearly at one and the same time, and in that event the lowermost individuals await the escape of those in front, which first push out through the neck of their earthy dwelling. One after the other, through one small hole, they all escape into the light of day. The actual method of hatching of *Stauronotus maroccanus*, as described by Künckel d'Herculais, is most interesting. According to him, an ampulla plays the principal part in the process. To escape from the capsule, the young Stauronotus puts into action an ampulla formed by the cutaneous membrane between the head and the prothorax. This ampulla is the more effective in being dilated by fluid from the body cavity, and is maintained in the swollen condition by the insect accumulating air in the crop.

Each capsule is closed with a well-fitting cover, and to detach it, six or seven of the young ones inside unite their efforts to push it off by help of their ampullæ. Nature has supplied these young ones with tools: sharp mandibles, powerful legs, furnished with spines and claws; but these seem not yet at their service, for a reason to be presently discovered.

We have by no means exhausted the important and

multifarious functions of this apparatus, the ampulla. It subsequently serves as a sort of reservoir, by aid of which the insect can diminish the bulk of other parts of the body, and thus after emergence from the capsule, penetrate the narrowest cracks in the soil, so as to reach the surface. As soon as it is there the young Stauronotus moults; the ampulla enables it to burst and to get rid of the skin in which it is enveloped. Freed of this pellicle, the young, no longer swaddled, can now make use of their limbs for walking and leaping, and have free use of their antennæ and buccal parts. At every moult the cervical ampulla reappears, and plays the leading *rôle* at these crises assigned to it.

But in the process of hatching of the Rocky Mountain Locust, Riley speaks of the feet as playing the principal part. By a continued series of undulating movements, and by the action of the sharp tip of the hind tibial spines, as also of the tarsal claws of all the legs, he finds the egg-shell is ruptured, and the nascent larva soon succeeds in working free therefrom and making its way to the light. Once on the surface of the ground, it rests for a moment, almost motionless. It is soft and limp, its members are still directed backward, and it is yet fettered in the very delicate film or pellicle, which must be cast before the newly-emerged creature can move with freedom. The skin begins to split, and in from one to five minutes from the time the insect arrives above ground, the process of extrication is complete. Pale and colourless when first having drawn itself out of this skin, the full-born larva is nevertheless at once capable of walking firmly on its legs, and even hops with agility, and an hour seldom elapses after the moult takes place ere its natural dark grey colouring is acquired.

However thin and delicate this pellicle discarded by the little animal on issuing from the egg may be, it doubtless affords much protection in the struggles of birth, and Riley points out the interesting fact that while, as we have just seen, it is shed within a very few minutes of the time when the animal reaches the free air, it is rarely shed if, from some cause or other, there is failure to get out of the soil, even though the young may be striving for days to effect an escape.

Post-embryonic Development.

Let us trace a migratory locust, *Schistocerca peregrina*, through its post-embryonic development; following Brongniart. Immediately on leaving the egg the young locust changes its skin, and is then of a green colour, but quickly becomes brown, and in twelve hours is black. Already at this early age the gregarious habit proclaims itself. In six days the individual experiences a second moult, after which it is black, mixed with white, and with a rosy streak on each side of the hind body. Generally six or eight days later the third ecdysis takes place; the general tint being the same, but the rose colour becomes more distinct. After eight days the fourth moult occurs; when the creature should be considered no longer a larva, but a nymph, for it has the first rudiments of wings; the position of the markings is the same, but the rose colour is altered to a citron yellow, and the line of the spiracles is marked with white. In ten days another moult is undergone, there is considerable increase in size, the yellow is brighter, and the prothorax more definitely speckled with white. In fifteen or twenty days the sixth ecdysis occurs, and the locust enters the perfect state. The large tegmina now present are marked with black, and the surface generally is rosy and bluish. Such is the colour in Algeria; yet sometimes the insects arrive from the South in the French colony reddish or yellowish in colour, those of the latter tint being, it is believed, older specimens of the red kind. We may recall an analogous series of colour-changes in the course of the individual development of some Phasmidæ—of the Phyllium group.

The fact of these changes of colour in Orthoptera during metamorphoses, and even after they have become adult, is important, not only from a physiological point of view, but as helping towards the determination of the number of species and the geographical distribution of the migratory locusts.

On an average, the Rocky Mountain Locust requires about seven weeks, from the time of hatching, to pass through its stages of growth; and the perfect form is attained through a series of five moults.

Apart from the colorational, and more minute structural changes which take place with each moult, the most striking change in the course of development is the growth of the wings. We examine the first instar of *Caloptenus spretus*, the young locust just emerged from the egg and colourless. In the second instar the chief difference is the development of colour; in the third there is plainly a slight development of the future organs of flight. After the third moult there is a great change; the instar then disclosed — the fourth — has undergone a considerable change in the wing-rudiments or wing-pads, which have become free and detached, the second pair being the larger, and outside the other pair. The fifth instar differs little, except in the increased size of the wing-pads. At the fifth and last moult the instar is the perfect insect, with full free wings, the thorax flattened, the colour different. The most pronounced changes, it will be observed, occurred at the third and fifth skin-sheddings, after each of which considerable difference in the form of the insect was presented.

Before each ecdysis the locust stops feeding for a while and remains motionless. The first three or larval skins are almost invariably shed on or near the ground, the young crowding together in some sheltered nook; for the last two or pupal moults they seem to prefer to fasten to some elevated object. Immediately after the operation the body is soft and colourless, as it was on leaving the egg.

Obviously these moults are most critical periods, and the last moult—from pupa to the winged—is the most critical, the most difficult of all. Clutching a grass-stem, or whatever be the chosen object, securely with its hind feet, which are drawn up under the body; the head downwards, if in the favourite position; motionless, with antennæ drawn down over the face, the whole attitude betokening helplessness,—thus the pupa awaits the swelling, and the ultimate splitting of the skin, when the struggle ensues to find release. Having succeeded at last in drawing itself out of its old "misfit," it, with unsteady gait, turns round and clambers up the side of the shrunken cast-off coat, and there rests while the body hardens, the crooked limbs straighten, and the wings dry, unfold, and expand, like the petals of some opening flower. The pale colours appertaining to moulting gradually give way to the natural tints, and a fresh bright locust eventually starts on a new career. Ravenous from long fast, by-and-by it joins its voracious comrades, and makes initial essay of its new jaws. During the helplessness that belongs to these crises, the unfortunate creature falls a victim to many enemies which otherwise would not molest it, and not unfrequently to the voracity of the more active individuals of its own kind.

CHAPTER VI.

LOCUSTS AND GRASSHOPPERS (ACRIDIIDÆ)—*continued*.

Locusts.

ONLY a few of the various kinds of grasshoppers belonging to this family Acridiidæ—of which nearly two thousand species are known—can be correctly denominated locusts. A locust is a species of grasshopper that occasionally becomes very destructive, and that moves about in swarms to seek fresh food. Some Acridiidæ greatly increase in numbers locally, and become very destructive—very often for one or two seasons only—and still more rarely migrate from place to place, and are then called locusts. The true migratory locusts are species that have the migratory instinct or disposition strongly developed, and that move over considerable distances in swarms, and inflict serious injuries. Of these there are but few species—less than a dozen, in fact—although we hear of their swarms in many parts of the world.

Perhaps the most important and widely distributed of them is *Pachytylus cinerascens*, which has extended its invasions over a great part of the Eastern hemisphere, from China to the Atlantic Ocean. It exists in many places in the Orient and the Asiatic Archipelago, and even in New Zealand, and is the commoner European migratory locust; its congener, *P. migratorius*, being much more restricted in distribution. *P. (œdaleus) marmoratus* has almost as wide a distribution in the Eastern hemisphere as *P. cinerascens*, but is more exclusively tropical. These locusts belong to the Oedipodides. This tribe does not include all the species of migratory locusts of the Old World; *Schistocerca peregrina* belongs to the tribe Acridiides. This handsome locust has a wide distribution. It is the chief species in North Africa, as in North-west India, and is probably the locust in Exodus. With this sole exception, the species of the genus Schistocerca are confined to the New World. *Schistocerca americana* is migratory to a small extent in the United States; and other species are migratory in South America. The

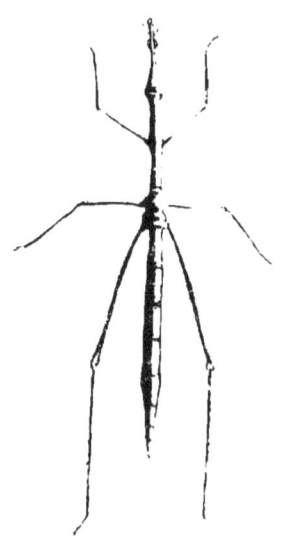

FIG. 21.—*Proscopia inæqualis*, which bears a great general resemblance to a Stick Insect.

genus Caloptenus also belongs to Acridiides. Several species of this genus are injurious in North America, but the migratory habit and great destructive power belong essentially to the Rocky Mountain Locust, the too famous *Caloptenus spretus*. While none other compares with it there in the vastness of its movements, or the injury which it inflicts, it, nevertheless, is a comparatively small, inconspicuous species, its slender brownish body seldom exceeding an inch and a quarter in length. We ascertain plainly from the above facts that the migratory species of Acridiidæ are not limited to any one genus or group of the family; and it is evident, therefore, we must look to something else than such anatomical characteristics or differences as distinguish the groups for the cause of the migratory instinct.

We should bear in mind this equally well established fact that locusts of the migratory species exist in countries without giving rise to swarms, or causing serious injuries; thus *P. cinerascens* is always present in different localities in Belgium, and does not give rise to swarms.

Occasionally individuals of these migratory species penetrate to our shores. In 1869 specimens of *Schistocerca peregrina* occurred in various parts of the country, having in all probability arrived by crossing the German Ocean, and *P. cinerascens* and *P. migratorius* have been met with; but Britain is now exempt from the ravages of locusts.

Migration of Locusts.

Insignificant individually, but mighty collectively, the migratory locusts fall upon a country liable to their visitations like a blight, for they appear suddenly on a spot in huge swarms, which, in the space of a few hours, clear off all the vegetation that can be eaten, leaving brown and bare that which all was green and flourishing. It is difficult for those who have not witnessed a serious invasion to fully conceive, or appreciate it. So great the proportions of the scourge, so vast their multitudes, their flight may be likened to a vast body of fleecy clouds, or, still more correctly, to an immense snowstorm, often extending from near the ground to a height that baffles the keenest eye to distinguish the insects in the upper stratum. It is a vast mass of animated specks glittering against the sun. On the horizon, they often appear as a heavy black cloud. So densely packed the brood, they occasionally even darken the air, intercepting the sun's rays for hours, and casting a checkered shade over the earth. Carruthers, in *Nature*, estimates a great flight of locusts that passed over the s.s. *Golconda*, when off the Great Hanish Islands in the Red Sea, in November 1889, at over two thousand square miles in extent; the number of insects he calculates to have been 24,420 billions;

and the weight of the mass 42,580 millions of tons, each locust weighing one-sixteenth of an ounce; and the ship of six thousand tons burden, he adds, must have made seven million voyages to carry this great host, even if packed together 111 times more closely than they were flying. Another, apparently a stronger, flight was seen going in the same direction next day. Other testimony goes to prove that such an estimate may be no exaggeration. According to official accounts of locusts in Cyprus, no fewer than 1600 million egg-cases were collected and made away with in 1881, up to the end of October; and by the end of the season the weight of the eggs collected and destroyed amounted to over 1300 tons.

From one part of the country to another the great hordes sweep, in search of pastures new, and leave ruin and devastation in their wake. The earth is entirely deprived of her green mantle, no green thing is left for beast or man. Famine is only a too probable consequence, and pestilence may follow from the decomposition of the bodies of the dead insects. The latter result is said on some occasions to have occurred from swarms falling into the sea, and being drowned, and being afterwards thrown upon the shore by the waves. Sometimes bodies of young locusts plunge into a stream and are entrapped there, the whole swarm being swamped in the river, there being no current to carry them away, and lie rotting in one huge mass for days. Fish die from the poisonous effects, and float on the surface, adding to the already existing mortality, and so powerful the effluvia produced, no one dare venture near to gaze on the scene of desolation.

Locust swarms do not visit the places that are subject to their visitations every year, but, as a rule, only after intervals of a considerable number of years. It has been satisfactorily ascertained that both in Algeria and North America, the noted locust years occur usually only at considerable intervals. There is, however, no certainty in the migrations, no law of periodicity governing destructive flights, these only occurring at irregular intervals. Nevertheless, it may be pointed out, the history of the most noted locust years, both in North America and in Europe, shows a tendency to their recurrence about every eleven years. From the respective areas of their most abundant development, the European and Asiatic, the African, and the American species swarm in exceptional years, to ravage adjacent regions in which they are not found permanently.

These interims between migrations seem at first unexplainable, for it would be supposed that as locusts are capable of excessive increase, when once they were established in any spot in large numbers, there would be a constant development of superfluous individuals which would have to migrate regularly, in order to procure food. The irregularity seems to depend on three things : that the multiplication of locusts is kept in check by parasitic insects ; that their eggs—which were supposed to be comparatively easily affected by climatic influences—may lie hidden in the soil for years and yet hatch out in a favourable season ; and that the disposition to migrate —though some locusts appear essentially migratory—is only effective when immense numbers of individuals are produced.

Every animal necessarily meets with checks of one kind or another to its undue multiplication, and the balance of power does not always lean to the side of the enemies of the locusts. During a year when the locusts are not numerous the abundance of the parasites may decline, and the bird destroyers of the locusts may greatly fall off in numbers ; so that the locusts get on the rising side of the scale, as it were, and for a time may increase rapidly, while the enemies are much inferior to them in numbers. If there should come a year when very few of the locusts hatch, then the next year the parasites will be greatly reduced, and if then large numbers of locusts from eggs that have been in abeyance should hatch out, the parasites will not be present in sufficient numbers to keep them in check ; and by the following year the increase in number of the locusts may be such as to give rise to a swarm.

There are Many Causes of Migration.

The explanation of migration to be gathered from the preceding remarks, is excessive multiplication. As a single explanation of the phenomenon no better can be offered; for this is evidently the immediate or special cause, and the others are mostly secondary, or but consequences of this one; for these movements cannot truly be ascribed to any single cause. Moreover, we come to recognize the fact that the several influences bearing on migration fall into two distinct categories, viz. immediate or special, and remote or general. That certain climatic conditions, those of heat and more than ordinary dryness, are in some way necessary, or most favourable to the excessive increase, and the origin of the migratory habit or instinct, may be maintained. Wherever locusts are largely and frequently developed, whether in the Eastern or Western Continent, we find either extensive deserts or vast treeless areas, and a climate arid or dry. These conditions are believed to develop irritating or uneasy sensations, which cause migration. But, plainly, there are other exciting causes than the impulse simply to fly. The immediate cause, increase, has been already considered. Annoyance from natural enemies probably often proves a valid cause, as no one who has witnessed the excessive abundance in which some of these at times prevail will deny. Tachina-flies especially, will follow the locusts in dense crowds, so thick that not one can rise from the ground without being pursued by several ; and there is no escape from persecution till the victim rises high in the air.

When food is lacking, whether through excessive multiplication of species, or through the droughts that are not uncommon in the locusts' native habitat, there must needs be the strongest incentive to change of place. Such is the case, under the same circumstances, with other animals normally non-migratory. That hunger will cause locusts to move from place to place in search of food is undoubtedly true, yet it would be a violent presumption to say that a swarm starts on the lengthy journey that they often undertake, in search of food. There seems thus still another factor in the problem, viz. instinctive impulse. It is more than probable there is a certain instinctive prompting to that which is best for the preservation of the species. Fresh breeding-grounds, away from the location of birth, seem desired. The evident disposition often manifested to go in a given direction in spite of contrary winds, or other potent obstacles, is not, to the same extent, susceptible of any other explanation. We have likewise to deal with that most remarkable fact, the return migration of locusts bred in the Temporary region to the land of their ancestors. Hunger or excessive multiplication, however important as causes of the migration from the Permanent breeding-grounds, seem to have little to do with this return migration, because the insects all depart, whether few or many, and they pass over great stretches of luxuriant vegetation. Prevailing winds do not govern them, dangerous obstacles do not stay them. The uncongenial climate of the Temporary region doubtless prompts them to get back to their more congenial native home, and we must allow a certain amount of instinctive guidance akin to that possessed by migratory birds.

Remarkable Manifestations of Instinct attend Migration.

The manifestations of instinct that attend migration are indeed remarkable. It is believed that when the locusts migrate, they do so in the direction taken by their predecessors, although several generations may elapse without a migration. Their flight is to a large extent dependent on the wind, and it is said when the air is calm and warm, and they are ready to move, they have the habit of making short flights, circling upward, apparently for the purpose of ascertaining the condition or direction of the upper currents, if they are favourable to bear them away. They support themselves on the wing probably with little muscular effort. Their body, we have seen, contains elastic air-sacs in connection with the tracheæ, which, constantly filled and refilled, float them up in the air, and at the time of flight, it may be presumed, the internal balloons have room for perfect expansion, as previously explained. Thus the insects spend but little muscular force in their aërial movements, and, instead of really flying as does a bird, are borne along chiefly by the wind. Suppose a swarm has risen from the ground on a clear sunny morning, and is being wafted straight on their course by favouring winds. Opposing winds suddenly prevent them from moving on in the direction they desire to go. Instantly they drop, and wait for a change. With a return of the wind to its original quarter, if during the warm part of the day, often, with wonderful unanimity, all again take wing, and fly off towards their intended destination.

An apparently inexplicable point in the natural history of the migratory locusts is their disappearance from a spot they have invaded. A swarm will arrive in a locality, deposit there a number of eggs, and then pass on. But after a season or two there will be few, or none of the species in the spot invaded. This appears to be partly owing to the want of food, and consequent mortality among the young after hatching; but in other cases they in turn migrate after growth—they disappear on those odd return travels towards lands from which their progenitors came.

"*Voetgangers;*" *Interesting Points in their Natural History.*

In South Africa it would seem that the movements of the migratory locusts frequently take place before the insects have acquired their wings. Mrs. Barber, in an account of "Locusts and Locust-birds," mentions this among many interesting traits of the South African species. These locusts manifest the gregarious disposition at a very early age, for the small family parties in which they at first appear, sunning themselves in the vicinity of the holes whence they have escaped from the ova, rapidly amalgamate, so that enormous numbers come together. Roosting in company upon shrubs and grasses, they denude the neighbourhood of its foliage. Their name, "Voetgangers"—given by the early Dutch settlers, and retained ever since—denotes their habit of travelling in flocks on foot. No sooner have they made their appearance from the earth, and have obtained a slight degree of strength, than they will at once take up a northern course, always towards lands in the interior of the continent from which their progenitors departed. Farmers and agriculturists adopt methods all but in vain to save their crops and pasture-lands on their approach in countless multitudes, when swarm after swarm has to be contended with. They then fill up the whole country; their numbers are overwhelming, and cannot be driven away. They take advantage of paths and roads, and many miles will be traversed in a day; they proceed by means of short quick leaps or hoppings rapidly repeated. Nothing can stay the "Voetgangers'" progress northward. Occasionally they march through towns and villages, and have been seen attempting to scale a stone wall, rather than be thwarted in their advance. Mountain ranges, forests, rivers, may intervene, diverting them for a while from their course; nevertheless, they ultimately succeed in continuing their journey to the interior.

The manner in which these wingless locusts occasionally cross broad rivers is surprising, and has some bearing on the vexed question of the possibility of winged locusts crossing the ocean. Mrs. Barber refers to an instance on the Vaal River in the spring of 1871, shortly after the discovery of the Diamond-fields. The country at the time was swarming with "Voetgangers;" every blade of grass was cleared off by them. One day a vast swarm appeared on the banks of the river, evidently in search of a spot for crossing, which they could not find, as the river was slightly flooded. For several days the locusts travelled up the stream; in the

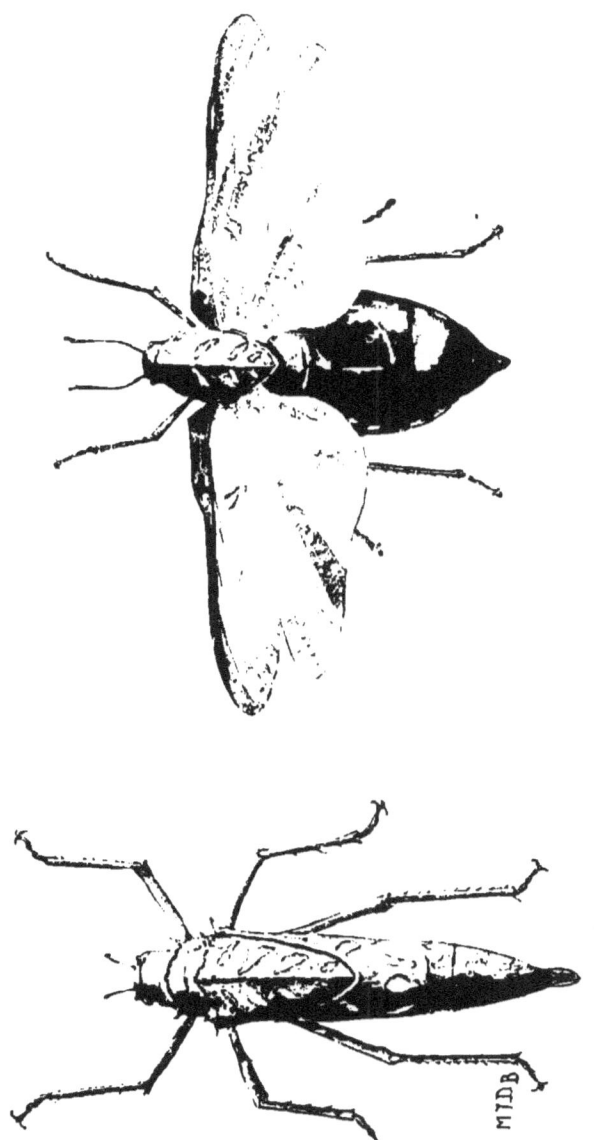

Fig. 22.—A very aberrant and beautiful Grasshopper (*Pucumora scutellaris*).

course of doing this they paused for some time at a sudden bend in the river where a number of rocks were cropping out, as if uncertain whether to risk a passage at this place. However, they at length passed on, as if with the hope of finding a better ford; in which apparently they were disappointed, for three days later they returned to the same bend, and there plunged in vast multitudes into the stream, where, assisted by a favourable current and the water-plants which grew upon the projecting rocks, they succeeded in effecting a crossing, though great numbers were drowned and carried away by the flooded river; but these casualties do not count when myriads are advancing.

Mrs. Barber adds that "Voetgangers" have been known to attempt the passage of the Orange River when it was several hundred yards in breadth, pouring their vast swarms into the flooded stream regardless of the consequences, until they became heaped up upon each other in large bodies. As the living mass in the water accumulated, some pieces of it were swept away by the strong current from the bank to which they were clinging, and as the individuals tightly grasped each other and held together, they became floating islands, the locusts continually hopping and creeping over each other as they drifted away. Whether any of the locust-islands chanced to reach the opposite bank is unknown; probably some of them were drifted on land again.

They are by no means rapid swimmers; but do not easily drown, their habit of continually changing places and hopping and creeping round and round upon each other being most advantageous as a means of preservation. It is a common practice for these young locusts to form a bridge over a moderately broad stream by plunging indiscriminately into it and holding on to each other, grappling like drowning men at sticks or straws, or, indeed, anything that comes within their reach, that will assist in floating them; meanwhile those from behind are eagerly pushing forward over the bodies of those that are already in the stream and hurrying on to the front, until at length by this process the opposite bank of the river is reached; so that a floating mass of living locusts is stretched across the stream, forming a bridge over which the whole swarm passes. In this ingenious fashion few, comparatively speaking, are drowned, because the same individuals do not remain in the stream during the whole of the time occupied by the swarm in crossing, the insects continually changing places with each other; those that are beneath are striving to reach the surface by climbing over others, whilst those above them are, in their turn, being forced below. Locusts are exceedingly tenacious of life, remaining under water for a length of time without injury. An apparently drowned locust will revive if exposed to the warm rays of the sun, should it by chance reach the bank or be cast ashore.

As a rule, the individuals who first acquire wings do not leave the general swarm, but remain with it until they all are mature, when the whole body will take wing together. In cases, however, where great swarms of different ages are mixed, the adults will suddenly arise and be off, northward, leaving their companions, the "Voetgangers," to trudge along behind until such time as they can follow.

By the middle of summer usually, the young have their wings, when they invariably all take flight and speedily disappear, travelling north, not a single specimen remaining behind in the Colony.

Distance to which Swarms may Migrate.

In the Old World the migratory locust is known to travel a distance of four hundred or five hundred miles into Central Europe from its permanent breeding-area in Asia; in North America the distance to which swarms may migrate in the course of a season may extend over a thousand miles.

The space over which a single flight may extend is not definitely known, but that it is considerable is certain. As the locusts rely chiefly upon the wind to bear them along, it depends upon the rate at which the wind moves, and the length of time they can remain in the air without descending. That as a rule swarms under favourable conditions arise in the morning, about eight to ten o'clock, after the dew is off, and the warmth of the sun is felt, and settle down again to eat as night approaches, by four or five in the afternoon, is well established: that is, they can sustain themselves an entire day in the air. The rate at which they travel is variously estimated at from three to fifteen or twenty miles per hour, determined by the velocity of the wind; a day's flight may therefore be estimated at from twenty to a hundred and fifty miles. But there are numerous facts which go to prove that a single flight may extend much farther than the longer distance here given: many persons believe the insects can remain in the air for days without coming down. While this is an extreme opinion, it seems undoubtedly true that they do sometimes continue their flights for more than a day; that they may, and do travel at night, has been recently clearly shown. Their general habit of alighting in the evening and resuming their journey next day after sunrise, if the weather is fine, together with the difficulty of observing them at night, have led to the general, but erroneous, impression that there are no exceptions to the rule.

There are good reasons for believing that the exceptions are much more numerous than might be supposed, that it is no uncommon thing for the insects to fly at night when the weather is warm and the wind favourable. This fact, in connection with the strong probability, nay, certainty, that swarms frequently fly at such a height as to be invisible in the daytime, will alone account for their repeated sudden and mysterious appearances in the morning, at other times in the afternoon of bright clear days, in localities, when nothing had been seen or heard of them along the line they had come. It is possible they prefer nights when the moon shines, though they are probably not confined to them : the warmth and wind being the influencing conditions. Obviously, this point is an important item in determining the possible distance to which single flights may extend. Flying two days and a night, say thirty hours, with a moderately strong and long-continued wind, they may pass over a distance of from four hundred and fifty to five hundred miles, and more, before alighting.

As evidence of the locusts' power of prolonged flight, the fact that they traverse seas of considerable width may be stated ; though the sea is undoubtedly often a source of destruction of swarms. They have been known to reach the Canary Isles from the African coast ; to come into Cyprus from the neighbouring coasts of Asia Minor ; to cross over the Red Sea. They have been seen in the Balearic Isles, having come from North Africa, and there are well-authenticated cases of their occurrence at sea. On November 2, 1865, a ship on the voyage from Bordeaux to Boston, when twelve hundred miles from the nearest land, was boarded by a swarm, the air being filled and the sails of the ship covered with them for two days. The species proved to be *Acridium* (*Schistocerca*) *peregrinum*. This is a most striking case, for locusts do not fly with rapidity, being, indeed, as we have seen, chiefly borne by the wind. It is possible some species may occasionally come down on the water at night, proceeding somewhat after the fashion of "Voetgangers" when crossing rivers as described by Mrs. Barber—in such masses as to buoy many without being submerged. An account of an occurrence of the kind may be read in Sir Hans Sloane's history of Jamaica, where it is stated that in 1649 locusts devastated the island of Teneriffe, that they were seen to come from Africa on the wind, and that, on the way over, they alighted on the water, in a heap as big as the largest ship, and the next day, having renewed their vigour under the influence of the sun, they took flight again and landed in clouds at Teneriffe. De Saussure says the great oceans are, as a rule, impassable to locusts, and that not a single specimen of the tribe Oedipodides has passed from the Old World to the New. Nevertheless, it may be that *Acridium peregrinum*, of the tribe Acridiides, was originally native to America, and migrated from there to the Old World.

Locusts' Enemies.

For dread of their hereditary enemies the locust-birds, the locusts will exert their extraordinary powers of flight; they are thought to travel day and night until completely exhausted, in the effort to elude these arch-enemies' pursuit and attack. But all in vain; sooner or later the birds overtake the locusts. There is the brown swallow-like locust-bird (*Glarcola Nordmanni*), the locust-eating stork (*Ciconia alba*), the grey mottled starling (*Dilophus carunculatus*), and others. All these are not only gregarious, but migratory as well, having no fixed habitat, but follow on the trail of the locusts in Africa. The first-named travels in multitudes second only to the locusts themselves, and are in every respect built for rapid flight. Coming up with their prey, according to their favourite mode, they "fall to" at their feast in its passage through the air. While they feed, at the same time they cut off with their broad beaks the legs and wings of the locusts they devour, which fall to the earth in myriads—a curious and novel sight. In stormy weather, when pursued by these birds, and they are unable to fly, the locusts seek shelter amongst the shrubs and grasses, creeping into them for concealment; but the birds descend in their train and hunt them up, taking advantage also of them in the morning when they are numbed with cold, and cannot rise.

It is upon the "Voetgangers" the various species of locust-birds depend for food for their young, not upon the winged imagoes, which are never stationary. Ascertaining to a certainty the locality where the locusts have laid their eggs, the birds there, for the time being, take up their abode. By the time the process of nest-building and incubation is accomplished, the young locusts have likewise made their appearance, so that there is abundance of food for the young birds. But occasionally dire calamity o'ertakes them—ere the latter are fledged the "Voetgangers" have started on their travels to the north. In most cases, when the locusts have arrived at maturity, and take wing, the locust-birds are ready to accompany them, and together, enemies and prey, they vanish on their wanderings.

Blackbirds, the Prairie hen (*Cupidonia cupido*) and quail (*Ortyx virginianus*), and the Plovers, are all efficient workers in the destruction of locusts in America; the Himalayan Black Bear, the skunk, squirrels, mice, frogs, and lizards, may be mentioned among other vertebrate animals more or less useful as locust destroyers. In Africa, in seasons of famine, when the hordes have swept over the earth, all creatures, including the human races of the South, take to a locust diet. But, excepting to the feathered tribes, it proves by no means wholesome food.

Invertebrate Enemies.

Insect and other minute enemies and parasites aid very materially in this destructive work. Many of the larger species of Ground beetles (Carabidæ) are most active pursuers and devourers of the locust. The swift-running and flying Tiger beetles (Cicindelidæ) have predaceous habits and similar tastes. One would hardly imagine that their stationary larvæ, living within cylindrical holes, and entrapping any unwary insects that may chance to come within reach of their formidable jaws, could succeed with such an active creature as the locust. Yet the young locusts fall victims to the larvæ; they are dragged to the bottom of the burrow, and devoured. Asilus flies are also particularly fond of them, pouncing upon and carrying them off to some nook where, unmolested, they can suck their juices. Digger Wasps catch them and paralyze them with their sting, and drag them into the holes they have dug in the soil, which are thus provisioned with food for the subsequent use of their larvæ. Scorpions make away with large numbers both of the "Voetgangers" and adults. They hunt by stealth; and on hot, sunny days, when the locusts are feeding, lie in wait and spring upon their prey, after which it is conveyed to their dwelling-places beneath the rocks. Trap-door spiders, when a swarm passes by, will suddenly rush from their den, seize a locust, and hurry it headlong down their gloomy hole, never again to see the light of day. Ants, mantidæ, dragon-flies, and others, doubtless do their part in helping to keep the locust in check.

Fastening themselves to it, under the wings, the Locust Mites suck the carcase to a dry shell, the dead bodies of the locusts sometimes almost covering plants, where they have taken hold of a leaf or stalk, and clasped it with a dead embrace ; while others fall to the ground to die.

The enemies of the locust so far mentioned either devour it bodily, suck its juices, or are parasitic upon it externally. There remain those which prey upon it internally, the most numerous and beneficial of which are the larvæ of the Tachina-flies. These attach their eggs to the parts of the body not easily reached by the jaws and legs of their victim, so as to prevent the egg from being detached. The locusts are attacked on the wing, and one haunted by a Tachina-fly makes frantic efforts to evade the enemy. But the fly persistently buzzes around, waiting her opportunity, and when the locust flies or hops, darts at it and attempts to fasten her egg under the wing or on the neck. Often she fails, but perseveres until usually her task is accomplished. The larvæ that hatch from these eggs eat into the body of the locust, and after living for a time on its fatty parts, issue and complete their metamorphoses elsewhere. A locust infested, though enfeebled, seldom dies till the maggots have left. So efficient is this parasite in America, frequently the ground is covered with locusts dead and dying from this cause; and in this manner whole swarms are rendered harmless. It will be remembered that the constant annoyances of these flies were suggested as at times one of the immediate causes of locust migration. In warm weather, the flies multiply rapidly, passing through their transformations in the course of a

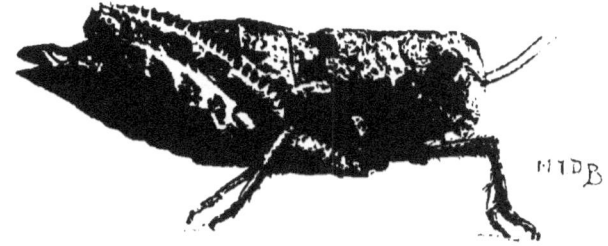

FIG. 23.—*Methone anderssoni*, specially remarkable for its complex organs for producing sound.

fortnight; but in the cooler seasons the development under ground is delayed, the winter being usually passed in the puparium.

Künckel d'Herculais finds that *Stauronotus maroccanus* and other Aridiidæ are followed on their devastating march by viviparous Diptera, which deposit their larvæ in the bodies of the locusts, the parasitism of these larvæ resulting in a sort of rachitic condition, answering to that produced by the Tachina-flies on the Rocky Mountain Locust. It is an odd and interesting fact that not a single member of the great family Ichneumonidæ—insects essentially parasitic—is known to attack the locust, either in Europe or America.

Some Species of Acridiidæ Present an Unusual Aspect.

The very large number of species of Acridiidæ have been recently arranged in nine tribes. Although the migratory locusts, just considered, and, indeed, the majority of this great host, are readily recognized from their family resemblance as belonging to the group, there are others that present an unusual aspect. Such is specially the case with the members of the tribes Proscopides, Tettigides, and Pneumorides, and with some of the wingless forms of the Oedipodides, a species of which we figure.

The tribe Proscopides (see Fig. 21) includes some of the most curious of the Acridiidæ. Of their metamorphoses, which are probably trifling, we are ignorant; of their habits we know little. Their colours are sober and dull, and Breitenbach describes the finding of some in abundance which he had overlooked for a long time, since they exactly resemble the withered vegetation amongst which they sit. When alarmed they seek safety with a lengthy and "lightning-like" leap. Their general resemblance to Phasmidæ, or Stick Insects, is striking.

But although the linear form and the very long body are common to both, this structure is due to the development of different parts in the two families. In Phasmidæ the prothorax is short, the mesothorax elongate, while in Proscopides the reverse is the case. The great length of head is very curious in these insects; but the mouth is not thereby brought nearer to the front, but is placed on the lower side of the head, quite close to the thorax. In most cases the sexes differ strongly from one another, and both usually are entirely apterous. But the genus Astroma displays a remarkable exception, and an almost unique condition of the organs of flight, the front wings being absent in each sex, while the female has rudiments of the hind pair which are wanting in the male.

Of the tribe Tryxalides, *Tryxalis pharaonis*, Klug, (see Fig. 19) is a handsome species, found in Sicily, Egypt, Algeria, Caucasus, and Andalusia. This insect approaches the Proscopides in form of the head and other characters. The tribe includes a great many species of grasshoppers. It and the Acridiides are the most numerous in species of the family; and to the latter belong most of the migratory locusts of the New World.

We come to the tribe Tettigides, a very extensive group of small Acridiidæ, remarkable for the shape of the pronotum, which is prolonged backwards as a hood reaching to, or beyond, the extremity of the body. In our British species this hood does not greatly modify the appearance of the insect, but in many exotic species, as in Xerophyllum, it assumes a grotesque form, so that the insects have no longer the appearance of Orthoptera, and when the genus Xerophyllum was for the first time described, it was mistaken for a bug. There is an immense variety of forms of this extraordinary prothoracic expansion. The odd *Cladonotus humbertianus* is a native of Ceylon, where it frequents sandy meadows after the fashion of our indigenous species of Tettix. Strange to say, some of these curious Tettigides, of the genus Scelimena, are amphibious in habit. They live on the borders of streams and ponds, and frequently dash on the water, and leap on their hind legs and flutter with their wings. In this way they strike the water without getting very wet; nor do they fear to enter the water, and leap about there; indeed it is said they prefer submerged plants as food. This singular habit has been observed in Ceylon and the Himalayas. The species have the hind legs provided with foliaceous dilatations probably suitable for swimming. These insects also frequent the little streams of water which trickle along the rocks, being often found clinging to a rock entirely immersed, as though they were enjoying a bath.

The tribe Pneumorides includes a small number of

species of very aberrant and remarkable grasshoppers, peculiar to South Africa. They are large, with short antennæ, and with the pronotum prolonged backward and hood-like. Unhappily, although amongst the most remarkable of insects, we are unacquainted with their habits. We give an illustration of the beautiful species *Pneumora scutellaris* (see Fig. 22). While the male is bountifully winged, the female must be necessarily sedentary, because of the imperfection of her alar system. From the form of the hind legs and their short length, it may be presumed that their leaping powers are null, or but slight. *P. scutellaris* is very remarkable from the difference in colour of the sexes. So brilliant is the female, she has been said to look as if "got up" for a fancy dress ball. She is of a bright green, with numerous marks or patches of pearly white, each of them invested with a disk of a mauve or magenta colour. Though the female is thus resplendent, her consort is of a modest, almost unornamented green. He is, however, furnished with a musical apparatus, by which he may be enabled to charm his gorgeous but dumb spouse. It consists of a series of fine ridges situated on the sides of the inflated abdomen, this part of the body having every appearance of being inflated and tense with the result of increasing the volume and quality of the sound.

The Pamphagides also chiefly inhabit Africa and the

arid regions on the shores of the Mediterranean Sea. Many are apterous forms, a circumstance that has, according to Saussure, exercised a marked influence on the geographical distribution of the species; but several species have well-developed wings. Some of the species are modified to a great extent for a desert life, and exhibit a great variety in the tint of the individuals in conformity with that of the soil on which they live.

Colour Difference Correlative with Locality.

To the tribe Oedipodides we have already referred as including most of the species of migratory locusts of the Old World. Some remarkable cases of colour-variation have been noted amongst the winged Oedipodides. In some species the hind wings may be either blue or red; it is thought that the latter is the tint natural in the species, and that the blue-coloured wings are a kind of albinism. But the most remarkable thing is that this wing-colour is correlative with locality. In the few localities in Europe where the blue variety of *Oe. variabilis* occurs, for example, not a single red specimen can be met with. Attention has been drawn to similar phenomena in other species in North America, and L. Brunner suggests that the phenomena in that country are related to climatic conditions.

He finds red-winged species most common in humid regions, yellow-winged in more or less arid districts, while the blue-winged forms are found chiefly in mountainous regions just between the dry and the wet conditions; and this same variation is observed among the representatives of the tribe in Mexico. So characteristic does this variation in colour of the wings appear, that he almost comes to the conclusion that an examination of a fair representative collection of these insects would be a sufficient index of the climate of the region from which they came.

Modifications for a Desert Life.

The Eremobiens, a subdivision of Oedipodides, include some of the most interesting forms of Acridiidæ. The most peculiar members of the group are some very large insects, specially modified for a sedentary and desert life. *Methone anderssoni*, found in the Karoo Desert of South Africa, and one of the largest of the Acridiidæ, is most noteworthy (see Fig. 23). This species is remarkable for its complex instruments for producing sound, and for the modification of its great hind legs, which have no saltatorial function, and but little, if any, power of locomotion, but act as parts of the sound-producing apparatus, and as agents for protecting the sides of the body. Both sexes are possessed of sound-producing organs, though the female is deficiently provided as compared with the male. On examining the first abdominal segment in either sex, a peculiar swelling is seen bearing two or three strongly raised chitinous folds, which being struck by some peg-like projections on the inner side of the base of the femur, a considerable noise is produced. Immediately behind the folds, there is a prominent striated surface: this is rubbed by some fine asperities on the inner part of the femur. These structures appear to be phonetic, at least in the male; in the female they appear to be somewhat less well developed than in the other sex. The male has still another phonetic structure. His tegmina, though rudimentary, are much longer than in the female, and their prolonged part is strongly ridged, over which moves the edge of the denticulate and serrate femur, giving rise to a louder and different sound. In the female there is nothing analogous to this; and there can be little doubt it is a sound-producing instrument peculiar to the male. In situation it approaches the musical apparatus of the males of the Stenobothri and other Acridiidæ. *Methone anderssoni* possesses large tympanal organs, which the small tegmina cover up.

In no other member of the Eremobiens, in no other Orthopteron, are the phonetic organs so complex as they are in the male of this insect. It would appear probable it has the power of producing two, if not more, distinct sounds.

In habit it is exceedingly sedentary, and apparently seeks safety in its protective resemblance and its stridulation, rather than by any attempt at flight. Its wings indeed are totally unfit for movement, being quite rudimentary, and the posterior legs seem equally unadapted for this purpose; they are enormous in breadth, dilated more than in other Acridiidæ. When Methone walks, it does so by means of its four anterior legs alone, on which it moves raised up as if on stilts, when probably the hind pair drag useless along the ground. In repose these odd hind legs are pressed close to the sides of the body, and the tibiæ are hid, the insect then—its colours being those of the desert sands— resembling, so closely as to be mistaken for, a clod of earth.

Among others modified to an extraordinary extent for their conditions of life, there is the curious *Trachypetra bufo*, another South African species. It lives among stones, and Trimen says it resembles with such precision the appearance of the stones that he had much difficulty to detect it. He noticed that in certain spots, often only a few square yards in extent, where the stones lying on the ground were darker, lighter, or more mottled than those generally prevalent, the individuals of the grasshopper varied similarly in colour in imitation of the stones. The differences found among individuals of the same species of desert Eremobiens are sometimes wonderfully great.

Two peculiar North American genera, Haldmanella and Brachystola, are known to represent this group in the New World. From its strange form and movements, *B. magna* has received the popular names of "buffalo-hopper" and "lubber-grasshopper." Almost equal in size to *Methone anderssoni*, it is not, like that insect, in correspondence with arid spots. It inhabits verdant prairies in temperate regions; and is more or less smooth of body, and green in tint, and rests concealed during the day under tufts of grass. Sound-producing instruments it has none, though there may be vestiges or rudiments of these.

Our native Acridiidæ belong chiefly to the genera Stenobothrus and Gomphocerus, whose musical instruments have been described. They are the little grasshoppers which are common in our fields, and well known to children for their saltatorial powers. The species of the genus Tettix are not uncommon. Besides these Acridiidæ, it will be remembered, three species of migratory locusts are occasionally met with in Britain.

CHAPTER VII.

GREEN GRASSHOPPERS (LOCUSTIDÆ).

THE family Locustidæ contains none of the true locusts. From the Acridiidæ we have seen these insects can be readily distinguished by their very long, delicate antennæ and four-jointed tarsi; and in other characters they differ essentially. Although as a rule provided with large wings, they have not the same power of flight as the Acridiidæ; it is believed there are no air vesicles connected with the tracheal system; and to this may be due the fact that none of the Locustidæ undertake those great migrations, so destructive to vegetation, that have rendered some of the foregoing family notorious. As a rule, they are less robust than Acridiidæ, their ocelli are considerably less perfect, and the head is often produced in front into a curious cone-shaped projection, the use of which is totally unknown, as in a large number of Conocephalides. One of their most characteristic features is the ovipositor, which frequently exceeds the length of the body. Not present in the newly-hatched

Locustid, the organ becomes gradually developed, and the completed structure is formed by the apposition of no less than six rods or pieces.

Life Histories.

Its mode of use is not always the same, some species depositing their eggs in the earth, and some place them

FIG. 24.—*Pterochroza ocellata*, its tegmina resembling leaves.

in parts of plants. Light soil, easily pierced, is chosen, and the apparatus thrust to a considerable depth into the ground. The British *Meconema varium* lays in the galls developed on oaks by the puncture of Cynips.

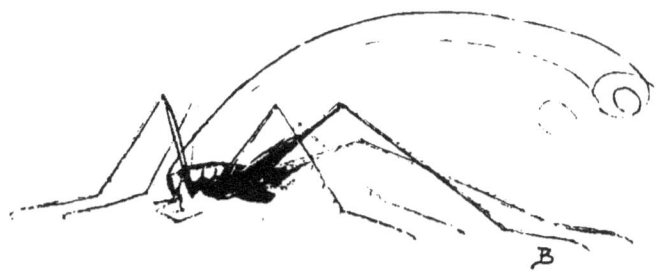

FIG. 25.—A cave-dweller (*Dolichopoda palpata*).

We find somewhat similar habits in North America, in *Xiphidium ensiferum*, a very common Locustid in some of the States, the silvery napiform galls produced by a species of Cecidomyia on the stems of willows being selected for the reception of the eggs. The insect thrusts its ensate ovipositor in between the imbricated scales, and leaves the eggs completely concealed, the overlapping edges of the scales springing back to their original positions as soon as the ovipositor is withdrawn. Some of these Cecidomyia galls are more or less fusiform, with flat, closely applied scales, which are far less well adapted than the more spacious interstices of the globular galls for the eggs. Between the closely appressed scales of the spindle-shaped galls, many are so flattened as to be incapable of developing; but the Locustids appear to show no preference for the globular galls. They seem, however, to prefer ovipositing, not in fresh galls, as might be supposed, but in blackened and weather-beaten specimens that have persisted through several seasons, probably because the scales are more easily forced apart.*

Since the imagines of *X. ensiferum* oviposit from the middle of August to September, and hatching is delayed until the following May or June, eight to nine months is therefore required for embryonic development, and the whole post-embryonic development must be passed through in three months. Development of the embryo begins in autumn, but a period of quiescence supervenes, due to cold weather.

Several of the congeners of *X. ensiferum* lay their eggs in the pith of twigs. Fissures in twigs and stems of trees and shrubs are chosen by many for the purpose. The eggs of *Microcentrus retinervis* constantly attract attention on account of their large size, their remarkable regularity of arrangement, and exposed position. They are deposited in one or two rows overlapping each other, upon twigs or margins of leaves. They are laid in autumn, during the day occasionally, but usually at night, and become more swollen as they approach the hatching period in spring. In the early part of May hatching occurs, the young insect undergoing a moult during the process. By the time emergence is completed,

* Embryology of this insect, Wheeler, *J. Morphol.*, viii. 1893.

it is a matter of wonder how the Microcentrum could so recently have been compressed into the comparatively small shell beside it. After a few minutes the little beings essay their first leaps, and soon commence to eat with avidity.

Including the moult in leaving the egg, they cast skins five times, acquiring wings at the fifth. Almost the first efforts of the liberated insect are directed to the task of eating up its out-grown and out-worn integument. Post-embryonic development takes about ten weeks. When first out of the pupal covering, the wings hang down on each side, limp and shapeless, but soon begin to dry and harden, and gradually are drawn into place, and assume their green colour.

Ears.

The Locustidæ resemble the Acridiidæ in the possession of special organs of hearing, or ears, but their situation is not the same as in Acridiidæ. In Locustidæ, in both sexes, a pair of ears usually occur near the proximal end of each tibia of the front pair of legs ; a tympanum, or a slit or crack opening into a cavity in which the tympanum is placed, is seen on each side of the legs. The ears may be divided into two principal kinds, according to the state of the tympanum, which is either exposed, or hid by a prominence of the integument. That these organs on the legs of Locustidæ are auditory organs has been ascertained beyond doubt. It is, in fact, only reasonable that insects provided with special sound-producing apparatus, as we shall presently see the Locustidæ are, should also be furnished with specialized ears.

A structure of a remarkable nature occurs in connection with the ears. At the posterior lateral angle of the prothorax, just over the insertion of the front leg, an open orifice is found; it is quite close to the prothoracic stigma, but is larger than it; it communicates with the one on the opposite side, and from them there extend processes along the anterior legs that narrow in passing the knee, and at last dilate in the tibiæ, so as to form longitudinal vesicles, one of which is in proximity to each tympanum of the ear. This prothoracic orifice, which exists in the two sexes, among nymphs, and among larvæ, and these leg-tracheæ, do not communicate with the tracheal system. Why the acoustic organs should require a supply of air other than that which could be obtained through the ordinary tracheal system, remains to be determined.

Although these tibial organs of Locustidæ have certainly the function of hearing, there is great difficulty in deciding as to the exact kind of sounds to which they are sensitive. Nor is it even understood how the sounds are transmitted to the nerves.

Musical Organs and Music.

The Locustidæ are famed for their musical powers. Among them, the stridulating apparatus, when present, is always situate on the base of the wing-covers. The left-hand tegmen has a roughened portion on the inner surface, serving as a file or bar, the right tegmen has a sharp edge on its inner margin ; and by rubbing the base of one wing-cover upon the other, and vibrating them rapidly, a musical sound, or stridulation, is produced. The extent of the delicate vibrating membrane of the wings which is brought into action is small, and when the tegmina are very rudimentary in size, it is these fields indispensable for stridulation that are preserved. With but few exceptions, the males alone are provided with organs of stridulation ; in the tribe Ephippigerides they exist in both sexes.

There is much variety in the structure. In this portion of the tegmina, the nervures not only differ in the male from those of the female, the arrangement of the nervures in the male is not symmetrical in the two tegmina. On the right tegmen is usually a small plate, or space, formed by a slender, tightly stretched, hyaline membrane, which, being put into vibration, probably increases the sound. In our large Green Grasshopper, *Locusta viridissima*, though the musical organs are by no means of a highly specialized character, each tegmen appears to act as a sounding-board.

Although in a large number of instances amply winged, the Locustidæ seem generally of a somewhat sedentary nature, availing themselves little of the wings for flight, using them rather as a kind of parachute, in descending from the trees to the ground. Moreover, they are chiefly nocturnal in their habits, though not entirely so. "These are the merry choristers," says Riley, speaking of American forms, "that make our woods and valleys ring with their pleasant songs during the evenings of late summer and early fall." The songs of the different species are very varied; indeed, each different species may ordinarily be distinguished by its peculiar note; but the study not infrequently presents some difficulties. *Locusta viridissima* produces a shrill stridulation, and sometimes chirps a little in the day. Bates tells us of one of these green grasshoppers, on the Lower Amazons, whose notes were the loudest and most extraordinary of any orthopterous insect he ever heard. Its native name, Tananá, is in allusion to its music. The natives keep it in small wicker cages for the sake of its song. The loud note of one could be heard from one end of the village to the other.

Katydids.

Amid the teeming exuberant insect orchestra of the American fields in autumn may be heard the notes of the Katydids, the most notorious of the singing Locustids —essentially American. There are several species of them—they belong, indeed, to several genera—but the song of all is supposed to suggest, more or less, the words of their popular name. Katy-did, katy-did, or, with variations, "O-she-did, katy-did-she-did," vociferates the garrulous "testy little dogmatist." Green leapers from leaf to leaf and from branch to branch, they might far more appropriately be called tree-vaulters than grasshoppers. Riley thus describes the music of the Angular-winged Katydid, *Microcentrum retinerve*, the commonest species in the Western and Southern States: "The first notes from this katydid are heard about the middle of July, and the species is in full song by the first of August. The wing-covers are partially opened by a sudden jerk, and the notes produced by the gradual closing of the same. The song consists of a series of from twenty-five to thirty raspings, as of a stiff quill drawn across a coarse file. There are about five of these raspings or trills per second, all alike, and with equal intervals, except the last two or three, which, with the closing of the wing-covers, run into each other.

The whole strongly recalls the slow turning of a child's wooden rattle, ending by a sudden jerk of the same."

From the setting of the sun till he begins to shed his rays in the East, these noisy choristers, during their most active period, will have it, with never an hour's remission, that "Katy did"—the species being so numerous that the sound as it comes from the woods is one prolonged rattling. Scudder states that these katydids sing both by day and night, but their day song differs from that of the night. "On a summer's day it is curious to observe these little creatures suddenly changing from the day to the night song at the mere passing of a cloud, and returning to the old note when the sky is clear. By imitating the two songs in the daytime, the grasshoppers can be made to respond to either at will; at night, they have but one note."

These Insects make quite Interesting Pets.

As with the species of the Amazon valley, they make interesting pets; they are even susceptible to domestication to a slight extent. Commendably neat in their ways, one of the most curious habits is an incessant polishing of the wings, legs, and antennae. Riley says *M. retinerve* brushes its face over with the front legs, just as a cat washes herself with her fore paws, and bestows

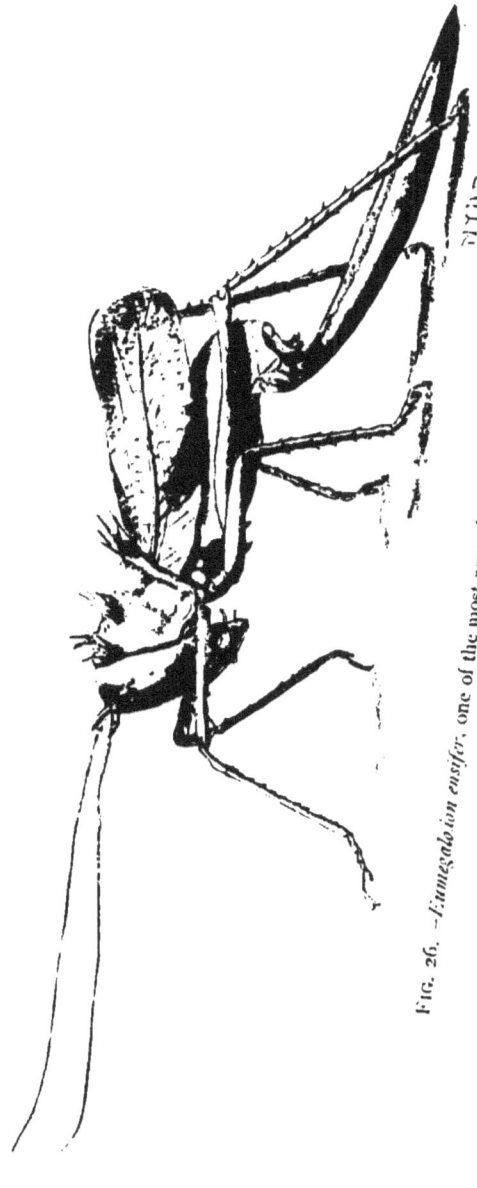

FIG. 26.—*Eumegalodon ensifer*, one of the most remarkable of the Locustidæ, from Java.

as much care on its long, graceful antennæ as many a maiden does upon her abundant tresses, the antennæ being drawn between the jaws and smoothed by the palpi, with evident satisfaction. But, in time, in his experience, confinement produces disastrous effects. He reared three successive broods in captivity, and, after the first year, the insects gradually deteriorated, so that the eggs of the third generation—the fourth spring—failed to hatch.

These katydids feed upon the foliage of the trees which they inhabit, but are rarely injurious to plants: Locustidæ are less exclusively herbivorous than the Acridiidæ are; many seem to partake of a mixed diet. A large number are believed to be entirely carnivorous, fewer to be solely phytophagous. It occasionally happens that they increase to large numbers in Europe, and in America in the case of a member of the genus Anabrus, which is sometimes destructive to crops.

The Tegmina resemble Leaves.

Many insects of this family are of a green colour, in assimilation to that of their habitat; the green of *Locusta viridissima* is wonderfully similar to that of the herbage amongst which it lives. The wing-covers in many present a most singular resemblance to leaves, colour, veining, shape, and appearance generally, being very leaf-like. Katydids have beautiful green and opaque fore-wings, like the leaves of trees, and display characteristic leaf-like veinings, which afford them protection from observation, and thus safety. There can be no doubt that the plant-like appearance of some of the Locustidae renders it most difficult to detect them in their native haunts. It is well known how these insects may attest their presence in considerable numbers in the immediate neighbourhood by their song, and yet it may be almost impossible to succeed in tracing the sound to its source, even in a single instance.

This resemblance of the tegmina to leaves is carried to the highest possible pitch of perfection in Locustidae. It is in the exotic genus Pterochroza, of South America, that the phenomenon is most marked (see Fig. 24). The tegmina in the species of this genus, in form and in tint, look exactly like withered leaves. In some of the species the wing-covers not only display the different shades of colour of dry leaves, but markings due to cryptogamic growths on leaves are reproduced. Not only this, transparent spots and tracks are present, like those on leaves due to the mining of insects. When settled, their tegmina closed, such insects are provided with perfect disguise.

Nor is this family entirely destitute of active means

of defence. For some, like the Phasmidæ, there is a more direct safeguard than mimicry. The Algerian *Eugaster guyonii*, if seized, ejects two jets of an orange-coloured caustic liquid from two pores situate on the sides of the mesosternum, and behind the coxæ of the front legs. A second, though feebler, discharge can be made, and sometimes a third; but then it has generally exhausted its store, and some time must elapse ere it has the defensive fluid again at command. This species may sometimes be heard making a low, brief sound, unlike ordinary forms of stridulation.

Some Locustidæ present a resemblance to Phasmidæ; this, however, has only been found in a few species. The long slender form and hind legs and long narrow tegmina give *Prochilus australis* a great resemblance to some of the winged Phasmidæ; while another Locustid, native to Australia, from its long and linear body and the entire absence of alar organs, looks like an apterous Walking-Stick. Saussure calls attention to the slender stick-like forms in the genus Peringueyella of South Africa. This genus belongs to the tribe Sagides; and, at first sight, one might almost mistake their extraordinary appearance for that of some Tryxalides.

Cave Dwellers.

There are other species particularly interesting for their strange habit of dwelling in caves. Whether found in the celebrated Mammoth and other caves of Kentucky, in New Zealand, or in Europe, they have a great general resemblance. Look at the enormously long antennæ and legs of *Dolichopoda palpata*, and its complete absence of wings (see Fig. 25).

Many of the largest and most singular forms of Locustidæ are distinguished for their constantly apterous condition, and these often have an unprepossessing aspect. The curious genus Anostostoma, with large head, and immensely developed mouth, armed with gigantic mandibles, certainly does not fall behind the rest in this respect.

Of the Curious Genus Deinacrida.

Allied to it are the interesting Wetas, a group inhabiting New Zealand. *Deinacrida heteracantha*, the great forest Weta, the "Weta-punga" of the New Zealand natives, is a remarkable insect. Formerly it was very abundant in the woods north of Auckland, but of late years has become extremely rare, the Maoris attributing its extermination to the introduced Norway rat. This giant species may measure over two and a half inches in length, and when the hind legs and antennæ are stretched out, it may be more than fourteen inches. Probably it subsists chiefly on the green leaves of trees and shrubs; Sir W. Buller remembers, in riding between Mangakahia and Whangarei, having caught a pair of Wetas on a low tree, where they seemed to be feeding on the young leaves. Although bulky and wingless, yet, as he tells us,* the insect climbs with agility, and is sometimes found on the topmost branches of lofty trees. When disturbed it produces a clicking noise, accompanied by a slow movement of the hind legs. "When taken it kicks or strikes backwards with its long hind legs, which are armed with double rows of sharp spurs; and unless dexterously seized will not fail to punish the offender's hand, the prick of its spurs causing an unpleasant stinging sensation." Killing them, so as not to injure the specimens, is difficult; and in one instance an attempt was made to drown a specimen in cold water, but it was found, after four days' immersion, as lively and active as ever.

A smaller species, *D. thoracica*, lives in decayed wood, into which it bores; and a third, *D. megacephala*, is characterized by a head and mandibles of exaggerated size in the male.

* *Zoologist*, 1867; *Trans. N. Zeal. Inst.*, 1894, vol. xxvii.

A perfectly new species (*D. broughi*), from Nelson, was reported in 1895. Apparently, it frequents dense forests, and lives by eating the heart of red-birch trees; it forms great tunnels, with enlargements or chambers, in the growing timber. Judging by the ways and doings of one of these Wetas in captivity, it is nocturnal in its habits; it became quite lively at night, and at times emitted a chattering kind of sound, which may be heard at night in the woods. It could bite fiercely, and, when excited, could hiss like an adder. Mr. Brough * found his captive Weta would eat nuts, and occasionally a little bark; but he could never induce it to feed by day. It could apparently, however, see perfectly well in the daylight.

As to the strange large insect *Schizodactylus monstrosus* (see Frontispiece), very little is known. The wings, which have their extremities much prolonged and curled, are a sufficiently remarkable feature; and it has no ocelli, and is believed to be wanting in ears. It is found in India, where it is said to be common in burrows by river banks, and has been recently reported as injuring young tobacco and other crops growing on high ground in Durbhunga, by cutting their roots. The local name given for the insect was *bherwa*.

* *Tr. N. Zeal. Inst.*, 1895.

Eumegalodonidæ.

Amongst the most remarkable of the Locustidæ are the species now forming the genus Eumegalodon, and the tribe Eumegalodonidæ. The ovipositor is long and sabre-shaped; the legs seem hardly in keeping with the heavy insect; the thighs indicate its addiction to climbing rather than to leaping exclusively (see Fig. 26). It was Brongniart who changed the name Megalodon into Eumegalodon; the genus Megalodon is placed by some naturalists among the curious Conocephalides.

LEPIDOPTERA

M

CHAPTER VIII.

FIG. 27.

The Fable.

WHAT measure could mete our indebtedness to Apuleius for his charming story—one of the loveliest myths of later antiquity—of the union of Cupid and Psyche? Apuleius introduces it into a playful satire on the follies and vices of the day, consisting of an elaborate romance, the "Metamorphoses," or "Golden Ass," so called probably from its affinity to the "Ass" of Lucian, the epithet "Golden" being added as a mark of admiration; but the best known, and by far the most beautiful, of the numerous episodes with which the work is interspersed, the fable of Psyche, is not derived from any source with which we are at present acquainted. Again the pretty tale, which, with seeming incongruity, is placed in the mouth of an old hag, will bear re-telling.

There lived in a certain city of Crete a king and queen, who had three fair daughters. The two elder were deemed worthy of the praises of mankind, and were early wedded to suitors befitting their rank; but the exquisite and inaccessible beauty of the youngest sister could neither be expressed nor sufficiently applauded by the poverty of human speech, and at length the crowds which collected in her honour from far and near, overcome with admiration, worshipped her as Venus herself. Venus's temples were abandoned, her ceremonies neglected, her images uncrowned, and her desolate altars defiled with ashes, while a girl was supplicated in her stead. Inflamed with jealousy, and raging high, that goddess commands her son of the flames and arrows to avenge the insult, and punish Psyche—for such was the name of the girl—by inspiring her with an infatuation for some despicable mortal, and Cupid—

> "Had still no thought but to do all her will,
> Nor cared to think if it were good or ill :
> So beautiful and pitiless he went.

Fig. 28.—The dragon's bride.

    And toward him still the blossomed fruit-trees leant,
    And after him the wind crept murmuring,
    And on the boughs the birds forgot to sing." *

But when he saw Psyche, he fell in love with her.

In the mean time her father, fearing—since, in spite of her renowned and glorious beauty, she remained unasked in marriage—that she had in some way incurred the displeasure of the gods, consulted the oracle of Apollo, and was directed to prepare his daughter for deadly nuptials: clothed in mourning garments, she must be led to a rock, there to become the bride of a dragon. Words fail to describe the anguish of her parents, who, day after day, delay the execution of the sentence; the whole city laments; but at last the sad procession sets out, and Psyche is left alone, weeping and trembling on the destined spot. Each moment may see the arrival of the monster; but lo! gentle breezes raise her, and waft her to valley, leaving her softly reclining on a bank of dewy grass.

When she opens her eyes, she espies a royal palace, full in every part of gold and silver and gems, precious beyond all price—a very abode for a god. Here she resides, waited upon by invisible servants; she has but to wish for refreshment and instantly banquets appear before her; a singer sings to her, but is unseen; an invisible musician plays; and she has a husband, who

* W. Morris, " Earthly Paradise."

invariably leaves her before the morning light; she has never beheld him, nor heard his name, and is threatened with dire calamity should she attempt to inquire concerning the person of her lover. But after a while Psyche's heart yearns towards her family, and she begs that she may receive the visits of her sisters. At first the sisters are content to embrace and rejoice over her whom they had believed to be dead, but soon the gorgeousness of her surroundings, her good fortune as compared with theirs, aroused within them feelings of envy, hatred, and malice; "I am not a woman," exclaimed one, "nor do I breathe if I do not hurl her headlong from such mighty possessions;" and they agree to unite to accomplish this. Repeatedly is Psyche warned to beware her of their perfidy. But Psyche is simple, of a loving and pliant disposition, no match for her sharp-witted, determined sisters. Simulating their base designs under an appearance of great affection, they entrap her into what amounts to a confession of ignorance of her husband, and induce her, by insinuation, to behold his countenance. Silently, at dead of night, she steals to the side of his couch, a lamp in her hand, and a razor, to cut off the head of the monster that she has been persuaded to expect. To her startled eyes no monster is revealed, but the beautiful god Cupid.

The razor drops from her powerless hand, and, terrified and shaking in every limb, she sinks fainting on the ground, until gradually the consciousness of the amazing beauty and radiance of the god revives her. There, at the foot of the bed, lie his bow and arrows, which Psyche grasps with eager curiosity, and drawing one of the arrows from the quiver, touches the point with the tip of her finger to try its sharpness, inadvertently piercing her flesh; and thus ignorant Psyche

FIG. 29.— *Hetaira esmeralda*, from Brazil : a clear-wing Butterfly.

voluntarily fell in love with Love. But now, suddenly, the lamp throws out a drop of hot oil on the right shoulder of the god. Cupid burnt, awakes, leaps from the bed, disengages himself from her clinging arms, and after reproaching her bitterly for her forfeited fidelity, springs with his pinions on high, and vanishes from the sight of his most unhappy wife.

Now follow Psyche's wanderings and trials. Inconsolable, knowing no rest, she is driven about from place to place seeking her beloved, her desire of finding him increasing with the difficulty of the search; then hope changing to despair, even of her own safety, she casts herself on the mercy of Venus: perchance she may meet him in the house of his mother. "At length," cries she, "have you thought proper to come and pay your respects to your mother-in-law! But take courage, for your reception will be such as a good mother-in-law ought to give;" thereupon her servants Solicitude and Sorrow scourge and otherwise

FIG. 30.—The Calliper Butterfly (*Charaxes kadenii*), from Java: sucking liquid from a muddy spot.

FIG. 31.—*Morpho menelaus*, from tropical America. Brilliant metallic blue and black.

torment the miserable girl, and, laughing and scoffing, she flies upon her, rending her garments, tearing her hair, and beating her, and subsequently sets her a series of impossible tasks, all of which Psyche accomplishes by extraneous aid; but there was no appeasing the ire of the enraged goddess. "There is one more labour, my dear, that you ought to perform," said she, with a sinister smile. To the realms of the Shades Psyche must go, to fetch a casket from Proserpine. Psyche is sensible that her last hour has come. Again, however, miraculous assistance intervenes, and she has almost fulfilled her embassy, and is returning from the infernal regions, when a rash curiosity seizes her to open the box, from which issues a deadly sleep, and she falls down motionless, in appearance a corpse.

Cupid could contain himself no longer, but flying swiftly to her, lovingly called her back to life, then, with rapid wing, mounted to the summit of heaven, and entreated mighty Jupiter on his behalf. A meeting of the Celestials is summoned, before whom Jupiter proclaims it to be his will that Cupid shall abide by his choice, and orders Mercury to convey Psyche to heaven; and as soon as she arrived, extending to her a cup of ambrosia, "Take this, Psyche," he said, "and be immortal; nor shall Cupid ever depart from thy embrace, but these nuptials of yours shall be perpetual;" and without delay, the wedding supper was served. And, by-and-by, a daughter was born to Cupid and Psyche, whom we call Pleasure.

The Origin.

While Apuleius cannot be credited with the invention of this poetically beautiful fable, his taste and feeling can hardly be too highly commended for this recognition of the capabilities of a wild flower of folk-lore. Doubtless the story is adapted from an ancient popular legend, of which traces are found in most lands. Everywhere the central situations in the tales are the same. The beloved may not be seen unveiled, his or her name may not be uttered; but the several taboos are always broken, and the pair are severed, sometimes for ever, or a reconciliation is effected, after protracted searchings and wanderings. We may believe, with Mr. Andrew Lang, that the myth arose out of a nuptial custom and law now forgotten—probably among some savages such a custom actually exists—the story being evolved to illustrate and enforce the law; and since these singular rules of etiquette appear to have been widely prevalent, there is nothing remarkable in the distribution of the myth among the most diverse races. The stories may have been separately invented in different lands, or may have been transmitted from people to people.

Into the myth Apuleius infuses an allegorical purport,

entirely in accordance with the Platonic philosophy.
Long before the beginning of the Roman Empire, the
Platonic myth of the fallen soul, who undergoes a series
of trials as a means of purification, to fit it for its mystic
union with Love Everlasting, seems to have taken a
great hold on the public mind in Greece, and was a well-
known and favourite subject with celebrated sculptors
and engravers of gems. Psyche signifies the *soul*, which
was at first symbolized as a butterfly, but like other
animal symbols, this of Psyche was gradually merged
into the human form, that of a slender girl, the wings
only being retained to mark its meaning. The gem
engravers exhibit infinite fertility of invention in depict-
ing scenes played by Love and the Soul. In some
devices, probably the oldest, Cupid appears alone,
chasing, tormenting, or caressing the butterfly; Psyche
in propriâ personâ he continually treats with harshness
and indignity, or the lovers meet, showing by tenderest
caresses delight at the reunion, or are united in marriage;
in sculpture they clasp each other in close embrace, as
in the group of the Capitol. We know that Apuleius
travelled in Greece, chiefly to acquire religious infor-
mation, and, clearly, these ideas and representations
were of a nature to afford subject-matter for his
imaginative genius to work upon in the construction
of his story.

Conspicuous Beauty and Abundance of the Symbols.

Than the Butterfly the world offers no illustration of the soul more striking, for it bursts in beauty on the wing from a dull, lowly chrysalis, its previous death-like stage, a lively image of the soul, freed and purified from material things.

Almost needless to say, butterflies are the most beautiful of all insects, in fact, for gorgeousness of attire, as a whole, they have scarcely a rival in the animal world; many may be likened in splendour to animated flowers. In the equatorial regions they constitute one of the most prominent and constant displays of animal life. Their abundance, their conspicuous hues, their size, as well as their peculiarities of shape and habits, all conspire to arrest the attention of the least observant of mortals. In the vicinity of old towns, both in the East and West, they are especially plentiful, and love to sport along roads and pathways in forests. Description can give no adequate idea of their rich and dazzling colours. Intense blues there are, satiny greens, gorgeous crimsons, fiery orange, golden yellows; not in small spots and patches, but in great masses, ofttimes contrasted by deep velvety black. We see wings spangled with metallic green, others literally glitter with spots and markings as of molten gold or silver;

some have changeable hues, like shot silk ; and several are as transparent as glass, as the *Hetaira esmeralda*, whose clear wings have but one spot of opaque colour, of a violet and rose hue (see Fig. 29).

The expanse of wing is not uncommonly six to eight inches in the Ornithoptera, or Bird-wings, so-called ; the largest, the most magnificent, the most perfect of butterflies, the wonder of the Eastern tropics. To the family Morphidæ appertain the largest and most splendid of the South American kinds. Their wings, often seven inches across, are usually of a brilliant metallic blue, and as the insect flies, the lustrous surface flashes in the sunlight, so that it is visible a quarter of a mile off. Slowly and majestically, the noble creatures sail through the forest glades, only flapping their wings at considerable intervals, or float at a good height, rarely descending nearer the ground than twenty feet, rendering them an almost unattainable prize in a forest country ; in the mountainous districts of New Granada and Ecuador they are captured with long nets, the collectors being sometimes let down by ropes over the edges of the precipices. A notion of the immense variety of butterfly life in the equatorial zone may be gained from the fact, that no fewer than seven hundred species actually exist within an hour's walk of Pará in Brazil ; while in Britain there are only sixty-six, and in the whole of Europe three hundred and ninety species.

Favourite Resorts.

Oddly enough, in the tropics, damp open places, especially river-banks and margins of pools, during the hottest part of the day, are favourite resorts of these children of the sun. Thirsty by nature, they congregate in absolutely countless numbers on the wet sand or gravel, to suck up the moisture. Often it is impossible to walk far without disturbing flocks of every variety of size and colour thus refreshing themselves, and at our approach they flutter up into the air from before our feet. In Nicaragua, groups may be observed of fine Swallow-tails (Papilio), greedily imbibing the water, and quivering their wings as they drink, and pretty blue Hair-streaks (Thecla), which,

FIG. 32.—*Cheritra jaffra*, brown with white tails, from Burmah.

SYMBOLS OF PSYCHE.

when they alight, have the strange habit of rubbing their wings together, keeping the curious tail-like appendages continually moving up and down. At many spots in tropical America sulphur-yellow and orange-coloured kinds (Callidryas) are very common, gathering in dense masses, their wings all held in the upright position, looking like bouquets on the ground, and when roused dissolving, as it were, into fountains of flowers; in association with white, or with brown and red butterflies, they represent, in the most deceptive manner, choice beds of flowers on the moist sand.

With few exceptions, these gaily tinted assemblies

FIG. 33.—Different females of the Malayan *Papilio memnon*.

which indulge in this sunshiny life are males, their spouses, which are more soberly dressed, and far less numerous, remaining hid within the forest shades, where every afternoon, as the sun goes down, the gaudy dandies join them.

Butterflies are occasionally migratory as well as

gregarious, and the hordes seem always to travel in one direction, to south-south-east, a fact throwing no light, rather the reverse, on these very unintelligible circumstances.

Twilight Fliers.

While some butterflies, usually of a dark colour, avoid the sun, by haunting the gloomiest recesses of the forest, others are crepuscular, issuing forth just after sunset, and flitting about in the dusk. Found both in the East India Isles and in America, these are true twilight fliers, appearing never to wander at night in the moonlight, or to enter lighted rooms as the Night-Moths, although, like the latter, they repose all day. So much for the universal influence of the warmth of the sun on the flight of butterflies!

A Quarrelsome Disposition.

Perhaps, also, we take too much for granted the mild, peaceable disposition, the defencelessness, the irrational character of the butterfly. A company will drive away and pursue for a short distance a large member of their kind which comes nigh their favourite resting- or feeding-places, assuring themselves that the intruder has really departed ere returning to their interrupted business. Doubtless they object to the addition of a guest at their feast whose appetite in comparison with theirs would be enormous. "Every one for himself" seems to be Nature's law : " finding 's keeping ; " and " union is strength."

Peculiarities of Highest Interest.

Closely allied to the Ornithoptera, or Bird-wings, is a group smaller in size, but equally brilliant in colour, the Papilios, one of the handsomest exotic species of which, the Malayan *Papilio memnon*, with black and blue wings, five inches in expanse, and with the hind pair rounded and gracefully scalloped, presents a peculiarity of highest interest—the remarkable variety in the form of the females. They may be divided into two groups—those which resemble the males in shape, though varying much, as butterflies often do, in colour. The second group is most extraordinary, and would never be supposed to be the same insect, since they differ entirely in colour and in the outline of the wings, the hind wings being elaborated into spoon-shaped tails, no trace of which is ever observed in the males, or in the ordinary form of the females. In shape and colouring these odd females, when flying, closely resemble another butterfly belonging to a different section of the same genus, *Papilio cöon*—a case of the wonderful phenomenon of mimicry. From some cause, the butterflies imitated seem exempt from the attacks of birds, and by imitating them the female of *Papilio memnon* also eludes persecution. It is, indeed, curious that both these distinct forms of female, the tailed and the tailless, are produced from the eggs of either form (see Fig. 33).

Of precisely the same nature as mimicry are those adaptations in which the insect is coloured and marked so as to represent the soil, or some vegetative object, the simulated appearance serving to conceal the creature from the prying eyes of enemies. The under sides of many butterflies, in all parts of the world, exhibit a deceptive

FIG. 34.—A Leaf-Butterfly (*Kallima inachis*), in flight and in repose.

resemblance to dead leaves; but the best example, perhaps the most perfect case of protective resemblance known, is to be found in the Indian butterfly *Kallima inachis*, and its Malayan ally, *Kallima paralekta*, both showy and conspicuous insects on the wing, but which no sooner alight than, as if by magic, they become invisible. Amid dried or dead leaves, on trees and bushes, it is their habit to rest, and in this position, with the wings tightly pressed together, they form a direct and finished representation of a leaf in some stage of decay. Colour, form, and habits all combine to produce this complete and marvellous disguise, and the protection it affords is shown by the abundance of individuals that possess it (see Fig. 34).

CHAPTER IX.

DAY-FLYING MOTHS.

FIG. 35.—*Urania Braziliensis*, migratory, from Brazil.

Lepidoptera Heterocera and Rhopalocera.

A POPULAR division of the Lepidoptera, or Scale-wings, in England, is into Butterflies and Moths, the former being termed Diurni, the latter Nocturni. In most continental languages one principal word serves for the two great Lepidopterous groups. Thus, *papillon* in French, may stand for either a butterfly or a moth; and they are distinguished respectively as *papillons de jour* and *papillons de nuit*.

But since, in fact, many of the species of the nocturnal Lepidoptera are day-fliers, and, *vice versâ*, not all the diurnal Lepidoptera fly by day, the habits of these insects do not seem to prove a good basis for separation, and in order to avoid the misconceptions produced by the terms "diurna" and "nocturna," Boisduval, a French entomologist, proposed to substitute Rhopalocera (club-horns) for the butterflies, and Heterocera (different-horns) for the moths.

At first glance few distinctions appear more happy than this—few classifications more natural. It was no sooner announced than it was recognized as a most convenient arrangement, and quickly came into general use. It is founded on the structure of the antennæ, often called feelers or horns, two long jointed organs situated in front of the head, between the eyes, which in this order are always conspicuous. A marked thickening towards the end almost universally characterizes the antennæ of the Rhopalocera. Such being the case, it is undoubtedly a character of primary importance. But a certain family of moths (Sphingidæ), by their antennæ thickening towards the end, though terminating suddenly in a point, bring the two groups into near relationship, and lessen their value; while the most interesting Castniidæ and Uraniidæ (of which more anon) so intimately connect them that these families have sorely perplexed systematists as to whether their rightful position was with the one group or with the other.

To be brief, our better knowledge than heretofore of the Lepidoptera gives rise to new views of the antennal structure, and makes plain the absence of any such an absolute difference. On this alone—on the clubbed or non-clubbed terminations, according to Boisduval, or on features which other entomologists deem more worthy of consideration — two primary divisions cannot be established.

In the same way, the stout spine or spring on the hind wings of moths is unsatisfactory as a classificatory basis. This spine is furnished on the under wings, at the costal base, and being received in a sort of socket beneath the superior pair, maintains them in a horizontal or somewhat deflexed position in repose, and is remarkably characteristic of the Heterocera. But all moths by no means possess it, while a butterfly is known armed with this apparatus.

In a word, though we may speak of Rhopalocerous and of Heterocerous characters, there is no one character which infallibly severs the two divisions, another instance of the fact that the naturalist has continually to face, the necessarily arbitrary nature of classification. The more intimate our knowledge of animal forms, past and present, becomes, the more our demarcations give way between all classificatory divisions, even from variety to kingdom. As we arrive at a true conception of the relations of animals, we realize the closer approach of the different groups, until we perceive an almost continuous chain. An observation of Kirby and Spence on this point: " It appears to be the opinion of most modern physiologists that the series of affinities in nature is a concatenation or continuous series ; and that though an hiatus is here and there observable, this has been caused either by the annihilation of some original group or species . . . or that the objects required to fill it up are still in existence but have not yet been discovered." Later-day naturalists find in these intermediate gradations, or transition states, such as I have indicated, and others, their strongest argument for the Darwinian doctrine of community of descent.

The Abnormal Collection of Pretty Insects Castniidæ.

Let us glance at the curious and abnormal collection of pretty insects, Castniidæ, which, in some respects, combines the characters of both Lepidopterous divisions, but in modern opinion has most affinities with the moths (see Fig. 36). Linnæus, and all the writers of the last century, regarded the species of the genus with which they were acquainted as butterflies, including them in the great group Papilio, on account of the clubbed structure of their antennæ. At the beginning of the present century, when this group was broken up, the genus Castnia was established; though Fabricius still retained it among the butterflies. But when the antennæ are carefully examined, they do not exhibit the real Rhopalocerous structure. In like manner, the Castnians differ from other groups of Heterocerous lepidoptera, in the complicated arrangement of the veins of the wings, and in various ways.

In general appearance they vary much, but, typically, they have large wings, with loose and remarkably large scales, and a position in repose deflexed or incumbent, being furnished with a wing-guide or guides; and the antennæ, though club-like at the tip, are generally long and more or less supple. All these characters are constant, and are Heterocerous characters. As a rule, the head is broad, and the body large and somewhat pointed. The Castnians resemble butterflies in this particular, in their evidently diurnal habits, as evinced by the brilliancy of their colours. They fly in open day, during the heat of the sun, with incredible rapidity, loving to rest wide-spread on the earth or on trunks of trees, and at certain hours pilfering the flowers of their sweets, and frequenting the inlets of thick forests, where they rest occasionally on the tree trunks, far above the ground. Numerous species are reckoned within their number, many of large size, and generally adorned with beautiful colours, a rich effect heightened by the metallic gloss of the prominent scales with which most of them are covered. In respect of colour the sexes may differ much.

Turning to their preparatory stages, the larvæ are endophytous, boring, with strong mandibles, the interior of stems and roots of cacti, of orchids, and other plants—a habit similar to that of the caterpillars of the Heterocerous Cossians and Sesians, but which, though found in butterflies, is very exceptional. Likewise they are provided with the ordinary horny piliferous spots or tubercles that characterize Heterocerous larvæ, and have a horny anal plate, whereas butterfly larvæ rarely possess these warts, but are frequently beset with close-shorn bristles, springing from the general surface, or from minute papillæ. In keeping with all Heterocerous borers, the pupæ are supplied with minute

FIG. 36.—*Castnia eudesmia*, from Chili.

spines on the hind borders of the abdominal joints, affording the pupa power of executing the same manœuvres as the chrysalid of Sesia, enabling it to move in the tunnel bored in the tree, and assisting it out of its cocoon.

The Castniidæ are essentially proper to the warm equatorial regions; their geographical range, in fact, extends only to Mexico and Central and Southern America, while they find their greatest development in Central America and Brazil. The few Castnioides, or species of Megathymus, known, inhabit the southern portion of North America, hailing from the Southern States, from Florida, and from Arizona. The genus Synemon appears to represent the Castniidæ in the vast continent of Australia.

The Yucca Borer.

Of these aberrant forms *Megathymus yuccæ*, the Yucca Borer, is one of the most interesting. Although placed with the Castniidæ, none the less it has given great trouble to systematists, having been bandied from the butterflies to the moths; and, it must be owned, some still regard it as a genuine butterfly. This species is common in the Gulf States of America over extended regions, where its larva commits serious depredations, of the nature that its popular name implies.

Fig. 37.—Yucca borer (*Megathymus yuccæ*), in flight and in repose; from the United States.

It is a sad-coloured moth, and rests with wings elevated—thus differing from the typical Castniidæ—its antennæ generally directed forwards; also in smaller wings, in smaller, closer scales, in unarmed hind-wings, and in stiffer, relatively shorter, antennæ, it diverges in character from the Castniidæ (see Fig. 37). Its flight, which is diurnal, is an extremely rapid darting motion as it passes from plant to plant, principally in open spots; though easily startled, it settles at no great height. During April and May, and earlier, it may often be seen in the morning where the yuccas abound, darting hastily about after its customary fashion, on laying thoughts intent; and as it pauses for a few seconds at one place, it fastens an egg to some portion of a leaf. The eggs are laid singly, though more than one may be put on the same leaf. They are subconical, smooth, and broader than high, pale green at first, changing to buff-yellow or brown. From the egg shortly hatches the larva—a reddish-brown creature with pitchy-black head—which shelters itself in a web between some of the young terminal leaves. Usually it starts proceedings near the tip of a leaf, working gradually downwards, eating the while, and rolling and shrivelling the blade as it goes. It lives thus among the leaves till about one-fourth grown, when it enters the trunk, commencing the devastation for which it is famed. Along the axis the trunk becomes bored and tunnelled out into a cylindrical burrow, wherein the larva makes its home, extending often two or more feet below the ground, and at its upper end lined with silk, generally intermingled with a white, glistening, powdery material, soapy to touch, and analogous with that of Hymenopterous and many Homopterous larvae. At what stage of larval development this powder is secreted is not known, but the full-grown larva is always covered with it more or less copiously, and doubtless it protects the invader against the mucilaginous liquor which the yucca freely exudes on maceration. How many larval moults occur has not been ascertained, but the insect continues in larva till subsequent late winter or spring, and during the coldest weather probably lies in a semi-torpid condition at the bottom of the burrow.

The funnel-like tube outside the burrow, made by the twisting and webbing together of the tender leaves when partially devoured, is quite characteristic of the larva of Yuccæ. The tube is, indeed, built and extended often several inches beyond the trunk or stem. From it the builder, especially when young, emerges to feed, and the small amount of matter besides silk used in its construction—the remnants of leaves and such-like substances—have been obtained and worked into the exterior from the outside. Pupation generally takes place at the top of the burrow, just below the funnel-like projection, but without the preparation of a well-formed cocoon. The pupa is of a brown-black, turning paler on the abdomen; all its members are distinctly defined, and, like the mature larva, it is more or less densely covered with a white powdery bloom. In due time the pupa skin is rent, and the imago issues.

The Brilliant Uraniidæ.

It has long been a problem with systematic writers what is the true situation in nature of the highly interesting group of insects Uraniidæ. Linnæus regarded the more typical species as butterflies, placing them in his great group called Papilio, containing the whole of the day-flying Lepidoptera; Fabricius, who divided the Lepidoptera into genera, even placed the genus Urania at the head of the order, followed by the other genera of butterflies. It may be mentioned that Urania may now be considered as the type of the family Uraniidæ, the genus Urania, in more recent times, having been split up into smaller genera. All detail must be omitted of the positions assigned to different members of the group from time to time. The day-flying habits of the insects, together with their airy forms and the extraordinary brilliancy of their colours, naturally led to their being at first classed among the Rhopalocera, or true Butterflies, but later acquaintance with their transformations proves them to belong to the Heterocerous division of the order.

Bosiduval describes one as "ce magnifique Lépidoptère, le plus beau de la création." They are among the most richly ornamented Lepidoptera of that very brilliant order. It would be difficult for art to effectually represent the changeable and resplendent golden green of the numerous bars, which contrasts with the velvety black of the wings, and varies with every change of light. It is these splendid tints of green that render them, perhaps, the most charming insects that exist, and has caused them to be named Emeralds; sometimes also they are called Pages. The posterior wings are prolonged into a single elegant pointed tail, closely resembling that appendage in many swallow-tailed butterflies; or there may be present at the hinder extremity of the wings no fewer than three distinct tails, that most remote from the anal angle longer than the others. The typical species of these superb insects are natives of tropical America, where they fly so high, and with such amazing rapidity, that it is almost impossible to catch them, and the only way, therefore, to obtain good specimens is to rear the caterpillar. When at rest, the anterior wings are kept in a flat or horizontal position, though only slightly spread, one peculiarity, amongst many others, in which they resemble the nocturnal Lepidoptera.

Urania Boisduvalii and Others.

Urania boisduvalii, which inhabits Cuba, may be considered one of the most beautiful Lepidoptera known,

FIG. 38.—*Urania boisduvali*, green and velvety black, from Cuba.

It attains an expansion of wing of from four to four and a half inches, with an undulated rim, the hollows of which are more or less sparsely tipped with white; otherwise its colours are velvety black and green. While the black of the superior pair of wings is relieved with golden green transverse lines, and their under side is nearly all black, with transverse lines of a bluish green, on the black of the inferior wings we note a longitudinal broad discal green band, a mark easily distinguishing this beautiful species from all its congeners. Its head is small and black, having a golden green V in the middle; the thorax is green and black; the eye has a black coppery lustre (see Fig. 38).

On approaching from the sea any open sandy shore in the Isle of Cuba a copse wood is perceived, above the coral reefs, forming a close and nearly impenetrable belt, maybe ten or twenty yards wide, and composed of almost one kind of tree, of aspect strange to the European eye, the *Coccoloba uvifera*, the so-called Uvero of the Spaniards. Immediately behind this belt an immense variety of vegetation grows in the parched sand—seaside shrubs in plenty, and what not, in general festooned with the flowers of different lovely climbing plants, every object quivering, as it were, beneath the scorching sun—the plant of chief interest to us being that technically titled *Omphalea triandra*.

This, the Cob or Hog-nut of Jamaica, the Avellano of the Island of Cuba, belongs to the family Euphorbiaceæ. Sometimes it reaches the dimensions of a tree, fifteen feet high. The part that concerns us is the leaves— great, thick, heart-shaped things of leathery texture, and a scabrous surface, of a pale green. The young leaves, and the leaves of the young plants, although of the same texture and colour, are of different form, being deeply incised.

During the heat of the day, on the upper side of the mature entire leaves of this tree, often the caterpillar of *U. boisduvalii* may be discovered torpidly reposing, screened from the fierce rays of the sun within a thin, transparent, silky web. At night, no longer sluggish, it quits its cover, greedily stripping Omphalea of its foliage, so that trees are left with scarcely a single leaf; nor is it inactive in the daytime when disturbed, but can run about quickly, and shows little affinity to the caterpillars of other diurnal Lepidoptera, which usually have a slow motion.

In February, and the ensuing months of spring and summer, the perfect insect deposits its eggs on the tender incised leaves, laying them singly, and apparently rarely attaching more than two to the same leaf, gluing each egg to its destined leaf by alighting on it merely for a moment. They are pale green, sometimes turning to yellow, and spherical as a rule; and on the whole their surface is not smooth, but ornamental, for from the summit proceed a number of longitudinal ribs, the spaces between which are crossed at right angles by obsolete striæ. The young larva, just hatched, is of nearly the same pale green, with a yellowish head, but ere it reaches maturity its appearance undergoes considerable alteration. It is then about two inches long, and moderately hairy; and while the body varies in tint from a pale yellowish green to a flesh colour, the head is now red, irregularly sprinkled with some black spots, and the prothorax is of a velvety black, though sometimes white predominates. The head, be it said, is polished and sessile, and never set on the body by means of a narrow neck. But these larvæ differ much from each other in size, marking, and colour, more so than ordinarily occurs with larvæ of the same species.

Eventually the larva spins an oval cocoon of dirty yellow silk, of which the threads are so few, and so loose, as to allow the inmate to remain visible through the meshes; within the cocoon it changes to pupa. The chrysalis is not at all angular, and, moreover, reposes in a horizontal position. Yet it agrees with that of most diurnal Lepidoptera in being rather gaily coloured.

Soon from the lowly motionless pupa arises the aërial imago, whose flight, which is truly diurnal, is swift, always strong, and in starts. The interior of the island it does not seem to haunt, but may be found plenteously in gardens at a distance of two, and even three, leagues from the shore, sporting in the sunshine, and sipping from the flowers of odoriferous trees of small stature. But it is by far the most common near the sea, because there grows its favourite Omphalea. As a matter of fact, however, it prefers flitting about the leaves of *Coccoloba uvifera*, unless when employed in depositing its eggs. About midday, and in hot weather, it is addicted to soaring particularly high, and surmounts the tops of the tallest denizens of the forest, tending as winter comes on to relinquish such ambitious flight, since it alights at times on hedges, offering, of course, a better chance of securing specimens.

As the genus Omphalea is common in Brazil and Guiana, in all probability it affords pabulum to *Urania braziliensis* and *U. leilus*, species whose habitats are Brazil and Cayenne and Surinam, respectively; for, as MacLeay has remarked, "the minor natural groups of Lepidoptera often keep very constant to the same natural group of plants." *Omphalea triandra*, which is very widespread in Jamaica, is doubtless also the cradle of the Jamaican *Urania sloanus*. *U. fulgens*, native to Columbia, Central America, and Mexico, certainly lives on arborescent plants of this genus. In a word, the gorgeous Madagascar *Chrysiridia madagascariensis*, the species of the East India isles, and many more, may likewise feed on leaves of seaside Euphorbiaceæ. Captain King, in his "Narrative of a Survey of the Coasts of Australia," describes having found one of the insects in immense numbers about a grove of Pandanus trees, growing on the banks of a stream which empties itself into the sea near the extremity of Cape Grafton on the north-east coast of New Holland. But MacLeay has "little doubt this species flitted about the Pandani as *U. fernandinæ* (*boisduvalii*) does about Coccoloba, while its eggs and larvæ might have been found on the neighbouring Euphorbiaceæ."

Migration of Day-Flying Moths.

Remark the habit of these day-flying moths of performing migrations. The very beautiful *Urania fulgens* migrates annually, from east to west, in August and September, across the Isthmus of Panama. Flights have been observed by a naturalist in the Isle of Caripi, near Pará in the Brazils, at Pernambuco, at Rio Janeiro, and in the Southern States of America; but he saw them nowhere so abundant as on the Amazons. From early morning till nearly dark, the insects passed along the shore in amazing numbers, but most numerously in the evening, and mainly, from west to east. Swainson, speaking of *U. brazilicnsis*, a species almost the exact counterpart of *U. leilus*, describes having witnessed a host flying during the whole of a morning in June past Aqua Fria (Pernambuco), in a direction from north to south—not one deviating from this course, notwithstanding the flowers that were growing around; and though they flew near the ground, they mounted over every tree or other high object which lay in their path, and it was impossible to capture a single specimen, so rapid was their flight. For three or four days they continued to pass in this manner. On the occasion of the flights over the city of Panama, in some cases the insects are attracted into houses by the light, so as to almost fill the rooms. They are accompanied by goat-suckers at night, and during the day by swallows and swifts, which probably destroy large numbers. As will be gathered, these migrations of Urania, though regular, are confined to comparatively narrow limits in the tropics.

CHAPTER X.

THE CASE MOTHS (PSYCHIDÆ).

Strange and Abnormal Lepidoptera.

IT seems an incontrovertible fact in natural history that there is not a single character which has been used to distinguish any group of considerable extent from which some one or more of the members thereof may not depart. In that great division of the animal kingdom characterized by the possession of articulated limbs, many species are met which are entirely wanting in those organs; and, similarly, the secondary division of the Annulosa, marked by the presence of wings in the final state—the Ptilota of Aristotle—contains species that, throughout life, never acquire instruments of flight. Of wingless insects, indeed, examples might be drawn from most of the orders, and in the majority of cases, the females only are thus deprived. Rarely, however, both the great characteristics are absent. Yet certain moths do not possess articulated feet in the wingless state.

Consequently, if we took into consideration merely

the adult state of the females, this group must be regarded as among the most degraded instances of apiropodous insects. But such a conclusion cannot be maintained, as shown by an examination of the early stages of the moths, for these, we find, exhibit as high an amount of organization as those of any of the other insects appertaining to the order. The truth is, these females have become degenerate—very different from the creatures they once were. Their peculiarity consists in this, that whereas, as a whole, winged insects always undergo a gradual evolution of structure, by which ultimately legs and wings are developed, these individuals gradually lose their powers of evolution, and not only this, but suffer a process of deterioration, by which the limbs which they at first possessed diminish, and at length dwindle altogether away, until the animal becomes a mere short, inert, vermiform bag, having not only no distinct trace of legs and wings, but also the sense-organs, the antennæ, and the organs of the mouth are almost or entirely obliterated, and even the articulated condition of the body has almost disappeared. In these extreme forms it is hardly possible for the degeneration of the female to proceed farther, and in all doubtless the change has occupied an immense period.

Than these extraordinary moths, familiar to German entomologists under the name of Sackträgers, perhaps no more curious and interesting examples occur among the whole of the insect races; certainly, in structure of the female, and in habit, they are the strangest and most abnormal of all Lepidoptera. They belong to the Psychidæ, a portion of the remarkable silk-spinning family of the Bombycidæ, but offer many points which are distinct in themselves, and entitle them to rank, as recent lepidopterists agree, as a separate and well-defined tribe.

Their geographical distribution is extensive, since they are found in Europe, in North and South America, the West Indies and Mexico, in Northern India and Ceylon, in China, the South Sea Isles, and Australia, being most abundant in sub-tropical regions. Wonderfully few species are described as natives of the United States; while in California, unfortunately three have been discovered solely in the larval state, the more mature conditions of the species as yet eluding detection. But there, as in various other parts of the globe, probably greater numbers await the industry of observers.

Among English-speaking folk the common appellations for the moths originate in the same circumstance as the popular term in Germany; house-builders, sack-bearers, basket-carriers, basket-worms, case-moths, — by these names they pass in England, America, and Australia, on account of the singular habitations, or sacks, they weave for the well-being of the caterpillars, in the early stages of their growth. Through the whole of their larval life they carry the protecting structure about with them; and as regards the apterous female, she never leaves this home in which she dwelt while in larva— one of the oddest incidents in this odd economy—but reaching maturity, and bringing forth her young, dies at last, without once quitting her self-constructed prison.

She deposits her ova, an immense number, within the body of the case, closely enveloped in some species in a short silky down; and almost as soon as the larvæ are hatched, they force their way out of the puparium which served for the defence of the eggs, deserting their early abode, and going forth into the world to follow independent lives. Escaping in crowds from the lower end of the tube to some twig or leaf, they immediately commence to prepare for themselves each a separate case, arranged in every respect as the larger ones, even before they have taken food.

Young Sack-bearers at Home.

The caterpillars manifest marvellous ingenuity in the construction of their cases. Particles of wood or bark, leaves, sticks, straws, lichens, mosses, and other vegetable

THE CASE MOTHS (PSYCHIDÆ). 207

FIG. 39—Larva case (*Metura elongata*), from Sydney.

substances, form, among the different species, the outer covering or decorative fortification of the house; the interior is lined with soft silk, and interwoven silky threads likewise bind together the external fragments. In the building materials chosen, and their arrangement, *Metura elongata* is a most interesting architect (see Fig. 39). Strengthening the large, elongate ovate bag of silk, and worked into it irregularly, appear numerous rows of short sticks, rather distantly separated, and about half an inch long, generally speaking; but towards the lower end there are usually several sticks from one to four inches long, in the centre of which the lower end of the silken bag protrudes, free from sticks, and very flexible. It has a charming silky softness, and is of a grey, ash, or mouse colour. Of this beautiful tissue the upper or head extremity is also composed, forming a tube half an inch wide. In the case of the Lictor Moth (*Entometa ignobilis*), consisting of a cylindrical bundle of slender bits of straight twig, about an inch in length, the sticks, as in the previous instance, are fixed longitudinally by the whole inner side to the flexible silken lining; the title Lictor is suggested by the resemblance between the cases and the fasces, or bundle of rods, borne by the lictors of old before the consuls. But a third Australian, *Animula huebneri* by name, has a case covered externally with a vast number of very slender twigs affixed to it by one end only, the other being free. Here we perceive an admirable piece of instinct, the loose points of the twigs being always directed backwards, so that in walking they oppose no resistance to the progress of the caterpillar, which they would do were they attached in the opposite direction, or without method. Cases of considerable size are armed with large pieces of leaf in similar fashion. It will be observed that these twigs, or leaves, as the case may be, are arranged somewhat systematically, the base of those nearest the narrow extremity lying beneath those nearer the wider aperture, or mouth, showing the latter to have been added subsequently, proving, in fact, that the case is gradually manufactured in the direction of its mouth. Through this the larva habitually puts its head. It is thus easy enough for it to extend its dwelling n this direction while resident within it, though it necessarily exposes itself to a considerable extent in applying the twigs or leaves to the outside.

The leather-like case of *Animula herrichii* is of remarkable construction, in that the external surface is destitute of any extraneous matters.

As a larva grows, needing more accommodation, it splits the habitation at the sides, weaving into the opening portions of the vegetable substances selected, and adding to the exterior fresh pieces of stick, straw, or leaves, as it requires. So with Saunders' Case Moth, when any accident happens to the nest, the caterpillar, with incredible expedition, repairs the damage received, employing the same silky stuff to fill up the hole; and with a nicety so perfect that the severest scrutiny cannot detect what was the extent of the injury.

Under the protection, then, of the substantial and somewhat formidable case the larva lives. At each end there is an opening, and through the anterior one it emerges to feed and change its position. Commonly, it only protrudes the head and the first three or four segments of the body, or sufficient to use its six true legs for locomotion when feeding; and if wishful to remain quiet, it usually takes the precaution of fastening a portion of the edge of the aperture by fibres of silk temporarily to the branch upon which it is, that, if alarmed, it can suddenly recede completely into the case, very rapidly drawing in the flexible part after it, by means of its mandibles and forelegs, and contracting the aperture so as to exclude all enemies. Thus hid, it stays in security, suspended only by a few threads. Were the nature of the hanging, tight-closed, strong, tough sack unknown, it would never be suspected of containing an active, voracious larva. Exceedingly wary and timid are these insects in retreating at the approach of danger. On a desire for removal, the suspending threads are bitten off close to the case.

As long as the caterpillar is small, and the house of no great weight, it is borne nearly erect, but soon, as a rule, the incumbent mass lies flat, owing to increased weight, and is dragged along in that attitude. The

THE CASE MOTHS (PSYCHIDÆ).

abdominal and anal legs of the larva are furnished with a series of small points or hooks, with which it moves in the tube, laying hold of the interior of the lining, to which it can adhere with great pertinacity; so firm is the hold retained, it is impossible to remove the creature without injury.

Coming of Age.

Having attained full growth, and being about to change to pupa, the larva of *Metura saundersii* firmly fixes itself, by means of silken fibres, spun for the purpose, to a branch or trunk of a tree, or paling, drawing together and permanently closing the head opening. It reverses its position in the case, so that the head is where the tail used to be, pointed towards the posterior or unattached end, and envelopes itself in a soft silken cocoon, of a yellowish colour. Allowing itself to hang perpendicularly, head downwards, it awaits the pupal sleep.

From the facts just stated, it need hardly be said that, when the time arrives, the perfect insect emerges from the posterior portion of the tube. At this particular time the male pupa becomes endowed with the power of stretching out its segments, to enable it to work its way out of the extremity. Through the opening of the posterior end it pushes the anterior half of its length by a slight elongation and contraction of the body, which, with the assistance of a transverse series of minute sharp spines or hooks, directed backwards, on some of the segments, is in this way forced out head foremost, in like manner as the pupæ of the Goat Moths and the large Swifts are made to emerge from timber and the earth, when the moth is ready to escape. The pupæ are prevented from being thrust entirely out of the case by two strong anal hooks. After the issue of the imago, the segments remain in their stretched-out condition. Cases having belonged to males are often seen with the empty pupa skin sticking rather more than half out of the lower aperture, hanging head downwards, as left by the moth.

It will be observed, both in the present species and the Lictor Moth, as in others, that there are, as I have shown, in most of the cases, one or two pieces of twigs longer than those of which the remainder of the case is composed, and extending posteriorly some distance beyond the termination of the fabric. Possibly these may be intended to assist the male, on entering the perfect state, to effect his exit from the case, the twig affording foothold, and aiding the imago to draw his long abdomen out of the pupa skin.

The males of these moths are swift flyers of extraordinary activity, dashing themselves wildly, almost to pieces, among the branches of the trees. A fiery little creature has no sooner arisen from his pupal slumber than he begins his violent fluttering, and as the wings are delicate in structure, in many instances nearly transparent, his beauty has generally disappeared before the entomologist can secure him, and specimens in good order are rare in collections. With slight exception, we find no homogeneousness in the perfect state of the insects of this group, but much variation of form

FIG. 40.—Male and female *Metura elongata*.

presented by the different species. The general shape of the body varies from one greatly elongated, as in *Metura elongata*, in *Dappula tertia*, and *Oiketicus kirbii*, to a short and robust, as well as to a short and slender form. In like manner the wings vary from a long, narrow, and sharp-pointed wing, as in Metura, to a wing of short, broad, and ample proportions; and again, may either be densely squamose, or colourless, of beautiful hyaline texture, almost or completely destitute of scales or hairs. The antennæ may be deeply pectinated only at the base, in others they are feathered to the tip; and in the number of joints offer striking variations (see Fig. 40).

Probable Cause of Disappearance of Beauty.

But the males of nearly all Psychidæ are characterized by a uniform dull dark colour of a brown or grey tint; there is an almost total absence of bright colour or of pattern. Yet these moths are in nearly all cases day-flying. Probably the beauty of the males disappears when the females become degenerate, and the conditions which produced it are then at an end. In other species of Bombyces in which the degeneracy of the females is less complete, less pronounced is the attendant loss of colour by the males. The day-flying Bombyces, whose females retain full possession of their faculties, are remarkable for the brightness and beauty of their colours.*

Singular Rarity of Moths considering Abundance of Cases.

The larval cases of these moths are amongst the "common objects" in Australia, meeting the eye every-

---

* Poulton, "The Colours of Animals."

where suspended to trees and shrubs, such as the different kinds of Leptospermum, Melaleuca, etc., fixed by their anterior end, and swinging loose otherwise. When unusually abundant, so as to look like a good crop of some seed or fruit, the pendant berths are particularly conspicuous, and attract the attention of the least curious of mortals. The most striking examples of the group belong to *Metura saundersii*, whose cases are sometimes over five inches long ; those of the male are one-third smaller ; but if this kind far exceeds any of the others in size, the case of the Lictor Moth bears off the palm for excessive abundance. The latter species chiefly frequents the Eucalypti, or Gum-trees so-called, but may also be found plenteously on many others of diverse botanical characters.

Considering this abundance, the insects are singularly rare in the moth state ; not one case in a hundred will be found to produce a moth, owing, partly, to the destructive effects of attacks on the larvæ of Ichneumonideous and Dipterous parasites ; even the Lictor Moths are surprisingly difficult to procure. From the same cause, nothing is harder, nay, more nearly impossible, it may be mentioned, than to rear these creatures in confinement. The caterpillars of a species may be collected persistently for years, and watched with incessant care, and yet never reach the perfect stage. Hence there are already imperfectly known species of which the more mature conditions await discovery; and when success does attend our efforts at protection, many examples are probably observed of the depredations of the insidious parasites. Not that failure to attain perfection is always due to infestation of parasitic insects, as undoubtedly the somewhat ponderous houses of the larvæ render them to a high degree impervious to the onslaughts of insect enemies: the cause of death must be looked for elsewhere. Death usually occurs after the larva has undergone metamorphosis, the pupa gradually shrivelling up after assuming its proper form, nor can anything be done, apparently, to avert the calamity.

A Perpetual Prisoner.

To return to the Case Moths' metamorphoses. The female insect, as we have seen, unlike the male, is destined never to desert the larval home. For her no hour of emergence ever comes. When the pupa has slept the appointed time, the unwieldy and almost motionless moth feels little of the movement of oncoming life then experienced by her lithe and lively partner; the animal, still resident within the habitaculum formed by the larva, splits asunder the pupa skin, and her transformations are complete: in some, at least, of the species the female imago is continually enclosed in the pupa case. Here, therefore, we have an insect which in its adult state is for ever excluded from the light, and never even beholds its mate.

In many of the genera the female is of the most degenerate type—a mere bag, a grub-like thing, the head, thorax, and abdomen hardly distinguishable from each other; without limbs or sense organs; totally unprovided with wings, legs, antennæ, or eyes. In the pupa case of these forms no distinct trace of former organs can be made out; in others they appear in a very rudimentary condition; in others again, still more distinctly. Observe the amount of development in two forms already mentioned, *Oiketicus kirbii* and *Oiketicus saundersii*. We see the ordinary grub-like appearance in the first, the three great divisions of the body being scarcely defined, and the whole enclosed in a tough envelope. Here exist neither tongue, palpi, nor antennæ,* no wings, and all but obsolete unarticulated feet: the general colour of the body is brownish; the neck and anus are clothed with wool-like hairs. Turn to *Metura saundersii*, about one and three-quarter inches in length, and in diameter full half an inch; cylindrical; of a pale brownish cream-colour, the head and thoracic segments light brown, fleshy, and smooth, the terminal segments clothed all round with dense silky down, of a deeper colour than the rest of the body. The insect is apterous, her antennæ

* Or only the very slightest rudiments of these.

are minute, unjointed, scarcely visible to the naked eye; but the feet, though very short and thick, are well articulated. It is this structure of the legs which at once distinguishes this imago from the larva; in this respect also it differs from the female of *Oiketicus kirbii* (see Fig. 40).

Having filled the bottom of their puparium with their ova, packed in the down rubbed from their own body, these females do not long survive. The moth is then literally nothing but thin skin. Reduced to a shrivelled, dried, and scarcely animated morsel of this matter, she either presses herself through the opening of the case, or, exhausted, the last feeble flicker of life burnt out, expires within.

CHAPTER XI.

THE HAWK MOTHS (SPHINGIDÆ).

Few moths are so attractive to lepidopterists, indeed, to all lovers of Nature, as the Sphingidæ, or Hawk Moths, partly from the beauty of the specimens. This is an extensive group, and its members, if not surpassing every other family of Heterocera in size, in their speed and indefatigable flight are unequalled.

Leading Characteristics of this Favourite Group.

They are a highly organized and specialized assemblage. A plump, robust, yet, as a rule, a graceful body, an usually long, conic, cylindrical abdomen, and prismatic antennæ—these may be said to be their leading characteristics; and generally the wings are comparatively small and narrow, the forewings extending far beyond the hind pair, and rather pointed; the tongue, though variable, is often strong and long, much exceeding the length of the insect itself, but is sometimes obsolete. In habits they are diurnal or crepuscular, some flying in brilliant sunshine, and others—the majority—in the early evening, in the twilight and just at dusk; and certain species are nocturnal and fly to light.

The larvæ are as readily distinguished as the imagos. Conspicuous, green in colour, hairless, and smooth, many are furnished with a prominent rigid spine near the tail called the caudal horn, which is sometimes lost in the later stages and then replaced by a shining lenticular tubercle. At rest, they have a remarkable fashion of elevating the head and thoracic segments, and curving them somewhat in sigmoidal shape, while they support themselves by their four or six hind legs, in which posture they remain for hours together, immovably fixed; and it is supposed that this attitude, giving them a fancied resemblance to the Egyptian Sphinx, prompted the name that Linnæus bestowed. They pass a solitary existence on trees, shrubs, or low plants, and suffer much from Ichneumonidæ, the check that alone prevents some species from becoming very injurious. When full-grown they transform above ground in an imperfect cocoon among leaves, or go underground and pupate in a cell.

Macroglossinæ.

Within this family six sub-families are included, and the marvellously specialized condition of some of their structural characters is brimful of interest. Most of the Macroglossinæ bear a complete resemblance to humming-

FIG. 41.—A long proboscis (*Cocytius cluentius*).

birds, due to their large, expansile tuft of hair-scales at the extremity of the abdomen, and their custom of hovering over flowers while sucking the nectar with their long proboscs. That these insects really are birds the uninitiated determinedly believe. The natives of the Amazons, as Bates relates, think that the moth changes into the bird, just as the caterpillar changes to the moth; Bates himself several times shot *Acllopos titan* in mistake for a humming-bird, so close is the resemblance between them on the wing. The Bee Hawks (Hemaris) have transparent wings, the clear spaces on emergence from the pupa being thinly spread with glittering scales, which fall on the first occasion of flight. During the hottest hours of bright sunny days they make their appearance among the blossoms and regale on their sweets, probably like all the Macroglossinæ with entire wings, such as Lepisesia; but those with angulated wings fly also in the dusk of evening. Genera of the following tribe are likewise taken in the day. Because of this diurnal habit, and the general idea that it is the most specialized group, entomologists usually place the Macroglossinæ at the head of the great series Sphingidæ.

Chœrocampinæ and Ambulicinæ.

To the peculiar tapering, often retractile, form of the larva, the Chœrocampinæ, the so-called Elephant Hawks, owe both their English and Greek appellations; Chœrocampa, the name of the typical genus, means Hog caterpillar. These moths are chiefly remarkable for their power of swift and long-sustained flight. The whole sub-family throughout shows a tendency to bright colours and distinct shades and bands in wing maculation. The abdomen is rarely banded, and there are few sober grey forms represented ; on the contrary, particularly in America, some are of a most beautiful and graceful description. Certain genera, as in the Ambulicinæ, have the anal segment of the abdomen in the males expanded laterally, an aspect approaching that of Macroglossinæ in this respect.

Sphinginæ.

When we turn to the Sphinginæ, we find these, as a rule, sober grey or brown, the usually moderately narrow and pointed wings of even outer margin, not sinuate or angulated as in the Chœrocampinæ. For the broad shades or bands of the latter, we have here a maculation consisting either of simple, undulated, transverse lines, or of longitudinal, interspacial dashes. Sometimes we get a mottled surface of grey and black without distinct pattern ; the abdomen, too, is almost universally banded or spotted ; and by these peculiarities in marking, and the wing-form, species of this sub-family may be easily recognized. Often their size is extraordinary, for members of Cocytius, a South American genus, which includes the largest known Sphingidæ, expand as much as eight and nine inches across the wings. The length of their tongue calls for similar remark, and some represent the extreme of development in this direction (see Fig. 41). A tongue five or six inches long is nowise uncommon; that of a tropical African attains to seven and a half inches; the proboscis of *Cocytius cluentius* reaches nearly two inches more! and Dr. Wallace some years ago predicted with confidence the discovery of a Sphinx in Madagascar with a tongue even of eleven or twelve inches, which could reach the nectar in the largest flowers of *Angræcum sesquipedale*, a singular Madagascar orchid, whose immensely deep nectaries vary in length from ten to fourteen inches. At the other end of the scale are forms like *Ellema harrisii* of the United States, where the organ is obsolete, or a mere membraneous rudiment. Another distinguishing feature of a large number of species in this group consists in their tendency to spinose, or armed tibiæ and tarsi, which in the Chærocampinæ is barely indicated in Deilephila.

These insects, as well as the sub-family Smerinthinæ, pupate in the earth without silk, an event more dangerous to the individual than in the cocoon-making groups, for the roving larvæ, seeking shelter, may find the ground unsuitable, and fall ready victims to attack. Their flight in imago is crepuscular and nocturnal without exception. There is thus a correlation between habit and structure; and, as we have seen, the higher Macroglossinæ and Chœrocampinæ—which may have a surface pupation and use silken threads—tend to discard the ordinary habit of the Heterocera and become day-fliers.

FIG. 42.—Hawk moth (*Lophostethus dumolinii*), from Port Natal.

Manducinæ.

The sub-family Manducinæ has always attracted attention, with its type *Manduca atropos*, the well-known Death's Head Moth, which is famed for its peculiar coloration and the squeaking sound that it utters. The pattern and colours, the ringed abdomen, and the contrast between the fore- and hind-wings, ally this Old World group with the typical hawk moths.

Smerinthinæ.

At the foot of the series stand the Smerinthinæ, or Eyed Hawks, with dentated wings more or less, small retracted head and thorax short and broad, and antennæ slightly pectinated in the males; insects often exceedingly richly coloured. In their maxillæ, or proboscis, they offer a striking contrast to the Sphinginæ. It is short or wanting, and of course, minus a tongue, the moths are incapable of feeding. They are heavy and sluggish in motion; the wings are not built for rapid or sustained flight, consequent upon the poorly developed state of the thoracic muscles. But they fly to light, and in this way many are taken. Withal they are truly Sphingiform in larval and imaginal characters, and have the anal horn, but the insects are thoroughly bombiciform in habit and appearance (see Fig. 42).

One of the distinguishing characters of the Sphingidæ, the reader is aware, is their smooth, hairless larvæ. Strange to say, it is a received opinion that they are an outgrowth of a spinous or bristly haired larval group, the posterior spine, or caudal horn, of Sphingidæ being regarded as a remnant of a general spinous covering. Probably the horn is developed from one or more spines or bristles, the skin itself at the base of which has been prolonged, and stiffened by chitine.

In form and structure the Sphingidæ seem to be most closely related to the American group of the Ceratocampinæ. If we regard the larvæ of the latter, we see the anal horn become stouter and more developed through the series Dryocampa, Anisota, Citheronia; the spines are gradually lost as the caudal horn became variously formed. The Sphingidæ have become smooth, and show only a thoracic crest or the anal spine represented by a horn. Probably the Sphingidæ were cast off from the Bombyces or Spinners parallel with the Ceratocampinæ. That the Sphingidæ may have been evolved from the ancestors of the Ceratocampinæ we have reason to believe. The sub-family Smerinthinæ would seem to be descendants of the oldest forms of the Sphingidæ. The small and sunken head, the shape of the thorax, and the pectinate antennæ are probably low characters in the Hawk Moths, and recall the Bombyces; as also the subterranean pupation without silk, together with the nocturnal flight. The mode of flight of the Smerinthinæ, which is opposed to the characteristic manner of other hawk moths, and their rudimentary mouth-parts, which prevent them feeding in the perfect state, likewise show affinity to the Bombyces. It is therefore manifest that these two marked features of the hawk moths—that they feed freely and are highly specialized in relation to flowers—were wanting in their Bombyciform ancestors, and are still wanting in Smerinthus.

When we turn to the general distribution of these moths, we find that while the Old World possesses a somewhat fuller and better representation of the higher groups, in America the grey, moth-like Sphinginæ exist in the greater number of forms. The representation in the Old World is the more brilliant of the two, owing to the number of bright-coloured Chœrocampinæ of the Himalayan region. In respect of colour, America has the advantage, however, as regards the Smerinthinæ; both Calasymbolus and Paonias, for example, are extremely beautiful. North America is indeed peculiarly rich in species belonging to the more typical ocellated group; but the less typical unocellated genera are not well represented.

Sphingidæ have their Metropolis in the Tropics.

The habits of the Sphingidæ, it need hardly be said, render them quite unsuited to cold, rigorous climates. It is in the tropics, where they revel in warm sunshine and a luxuriant flora, that they have their metropolis—where they reach the highest development in kinds and numbers. The family is not represented in Iceland, nor in Labrador, in all probability; but they occur in Vancouver and in Newfoundland, and in Upper Canada the majority of those inhabiting the Middle States may be found. *Hemaris diffinis* extends further north than most other species.

Occasional Visitors from Sunnier Climes.

But, now and again, these colder latitudes receive visits from the species of the sunnier climes. To Europe in this way comes the celebrated Oleander Hawk Moth from Africa; and our *Chœrocampa celerio*, the Vine Hawk, is probably only a casual visitor with us. The summer winds bring like occasional welcome guests to the eastern portion of the States of America, from the West Indies, and Florida; among them the Blue and Green *Argeus labruscæ*, *Dilophonota ello*, commonly called the Wandering Hawk, and the Bee Hawks, *Aellopos titan* and *tantalus*, the last from the Antilles. *Triptogon lugubris*, *Dupo vitis* and *linnei*, and *Phlegethontius rustica* may be also mentioned as tropical species appearing irregularly, or as wind visitors in the north.

As wanderers over the face of the globe none can compete with the Chœrocampinæ; and as swift flyers they have no rivals, with their pointed bodies and beautiful clear-cut wings. *Chœrocampa celerio* has the greatest geographical range. Always rare with us, it is met with the vine everywhere throughout the warmer parts of the Old World, and not unfrequently is caught on board ship, out of sight of land. No species more admirably exemplifies the family renown for rapid and indefatigable flight.

CHAPTER XII.

THE DEATH'S HEAD MOTH.

FIG. 43.—The Death's Head Moth.

It is in Several Respects a Most Remarkable Species.

THE remarkable and beautiful Sphinge, the Death's Head Moth, attains an expansion of wing sometimes not far short of six inches, being therefore not only the largest of our indigenous Lepidoptera, but, with one exception, the largest insect in Europe. Extremely broad and thick of body, with a large head, and a short thick proboscis or tongue, it bears on the back of the thorax a conspicuous, well-defined, yellowish blotch, containing two round black spots and some dark grey clouds, not inaptly representing the face of a human skull, or Death's Head. The fore wings are thick and strong, and of a blackish grey, mottled with yellow and red; the hind ones of a rich brown-yellow barred with black, and with these wings striking aberrations, when they do take place, seem to be usually connected.

Similar qualities distinguish the larva. It also is large and handsome, solid and thick, and about five inches long, and, when at rest, in the habit of raising its anterior segments and drawing them back, in assuming the curious sphinx-like posture, so characteristic of the group. Its colour is green or yellow, sprinkled with numerous small black or purplish dots, and with seven broad oblique lateral stripes of a dull blue or violet, extending to the back, where they meet in an angle; and the twelfth segment carries the horn, which is rough, and strangely bent downwards, and then recurved again at the tip. But this is a most variable larva, for specimens are occasionally found of different shades of brown, with the stripes much less distinct, or even white, and sometimes the first three or four segments have broad whitish stripes and patches, and instead of the usual lateral stripes, a chain of brown diamond-shaped cross-bars.

In July and August it occurs in potato fields, busily engaged in devouring the leaves; but as it feeds during the night, and remains hidden low down on the stem throughout the day, it is not so easily discovered as its size would lead one to imagine. Formerly scarce in this country, since the cultivation of the potato it has increased considerably, though not to an extent sufficient to do any real harm. But in the event of the potato being destroyed by disease, it will attack other plants, and has been known to take to them so kindly as to refuse the potato when supplied with it. It also lives naturally on the tee-tree, the common jasmine, the deadly nightshade, woody nightshade, snowberry, dogwood, and various others of very dissimilar qualities. About the middle of August, as a rule, it becomes fullfed, when it retires into the ground to a considerable depth, and forms an oval cell, carefully smoothed inside, wherein to undergo its destined changes. Owing to delicacy of skin, the pupa is frequently injured when the potatoes are dug up in the autumn, and without precautions, disturbed pupæ almost invariably succumb before reaching the adult state.

The Moth has a Voice.

Its superior dimensions and bulk of body, and the singular markings on the thorax, which bear such a wonderful resemblance to a human skull, conspire to render this moth a most remarkable species, but it is still more striking and unique from the fact of possessing a voice, or the power of uttering a kind of shrill, plaintive, and mournful squeak, somewhat resembling that of a mouse—a peculiarity appertaining to only one other species of the family, not belonging to this genus (Manduca). While both sexes can produce the noise, and some individuals do so with the greatest readiness whenever touched or disturbed, nothing will induce others to make it, ever so faintly. The strange cry has been long known to naturalists, and the question of its origin has given rise to much discussion. Almost innumerable theories have been invented to account for this apparently simple phenomenon, and quite a literature of its own has accumulated round the subject. From Reaumur downwards, observer after observer has experimented with the view of ascertaining the exact seat of the sound. Some have attributed it to the same cause as in certain beetles, the friction of one organ against another, as the rubbing of the proboscis against the palpi, or the thorax against the first segment of the abdomen; and it is ascribed to the forcible expiration of air through the tongue or trunk ; and it is said that bubbles have been seen to form upon the tongue when the moth had been induced to squeak under water. A large dome-shaped cavity has been shown to exist in the head, which, by the alternate action of elevating and depressing muscles, is caused to act as a bellows, and probably inhales, as well as exhales, air through a narrow slit-like aperture, leading to the proboscis. That the air expelled enters the cavity from behind, as has been supposed, is hardly likely, for the posterior opening is small, if not often altogether aborted, while the animal can still squeak if its abdomen be removed. At the narrow slit-like opening the note is formed, the sound being modified by passage through the proboscis tube.

But it has been satisfactorily proved that the pupa has the power of squeaking like the moth shortly before emergence, and, it must be confessed, it is difficult to understand how the methods of production suggested can operate in this case. Strange to say, the larva has also a voice, of a totally different nature however, being a peculiar grating or crackling noise, that may be compared to the snap that accompanies an electric spark, and sometimes the noise is repeated in rapid succession, resembling that occasioned by the winding up of a watch. Cottagers finding the caterpillar have described it, not inaptly, as biting its teeth at them. There is no doubt the sound is of a defensive character, and is made when the animal is irritated or disturbed. It appears to result from a lateral action of the large mandibles or jaws, which are furnished on their outer surface with some minute prominences; and when one jaw is outside, and passing over the other, it is momentarily arrested by the prominence of the latter, and falls sharply against its outer surface towards its base, the sudden jerk and collision between the two hard chitinous substances probably causing the sound.

An Object of Alarm to the Superstitious.

It is not surprising that a creature invested with so many startling attributes should be the object of superstitious beliefs and alarm among ignorant country folk. That it is nocturnal in habit, concealing itself in some obscure spot during the day, and appearing on the wing only in the morning and evening twilight, serves but to intensify the unfounded fear. In Eastern Europe, where some years it is extremely abundant—so much so as to enter houses, and at times extinguish the lights—it is regarded with horror as an evil omen, a forerunner of war, pestilence, famine, and death to man and beast. They call it the Death's Head Phantom and Wandering Death's Bird in German Poland, convincing

evidence of the light in which it is held. To these
fertile imaginations, the grim features stamped thereon
represent the head of a perfect skeleton, its cry becomes
the moan of anguish, or grief, or of a child; the very
brilliancy of its eyes typifies the fiery element whence it
came, for they implicitly believe it to be a messenger of
evil spirits. Once, on its plentiful occurrence in Brittany,
a country prone to superstition, it created the greatest
trepidation among the inhabitants, its appearance co-
inciding with a disastrous epidemic, which they charged
it with bringing, or, at least, that it came to announce
the fatal malady. An idea prevails, among the Creoles
especially, that it is very dangerous, in that the dust
cast from its wings in flying through a room will blind
those in whose eyes it falls, and thus it is driven forth
by every means. Even some parts of England have
the saying that the moth is in collusion with witches,
and whispers in their ear the name of the person for
whom the tomb is about to open.

As a "Bee-robber."

Perhaps owing to its habit of flying late at night, it
is not quite clear whether or no it gathers food from
flowers. It is seen hovering at flowers, it is said,
though rarely, but has never been caught while so

employed, and the shortness of its tongue compels us to feel rather doubtful of its capability in this direction; some suppose it to suck not from flowers, but the exuding sap of trees. However, in common with the predilection of most Lepidoptera, it is strongly attracted by honey, and apparently appreciates it in larger quantities than flowers supply, being well known to enter bee-hives when it gets a chance. It has been found trying to gain access to the hives, disposing of all doubts as to the habit, and once inside, its task must be to effect the enjoyment of its meal in peace. This inoffensive creature, its thick skin and downy covering notwithstanding, seems absolutely incapable of resisting its armed assailants. Its huge size may scare the bees, and its stridulous voice has been thought to arrest and control the hostility of this irritable race, in a manner similar to that produced by the song of their queen. If this conjecture be correct, we may ascribe the rare instances in which the moth has been securely fastened down inside the hive to natural death of the moth therein, the bees, being unable to eject so bulky an object, having taken the precaution to embalm its body with the glutinous substance called propolis; indeed, this circumstance could not well occur to a living moth, ineffectual though it be, unless it were completely stupefied by gorging on the honey. In the south of Europe, in some years, it becomes very injurious in this way to bee-hives. The construction of modern bee-hives keeps it out, and when the old-style hives are used, an efficacious remedy lies to hand, by covering the opening into the hive with wire grating, fine enough to prevent the entrance of insects larger than its rightful inmates. According to some Continental apiarians, the bees are aware of their liability to the intrusions of the moth, and when located in the old-fashioned hive, erect a kind of fortification at the portal, including a narrow and turned passage, through which it would be impossible for the moth to squeeze, while it is equally powerless to force its way through by biting.

Essentially a creature of the night, the Death's Head Moth can hardly be roused into animation in the day; even by pinching, and throwing it into the air, it can only be induced, and that in sluggish fashion, to flutter the shortest distance. But on the wing, at night, all is changed, for its power and endurance seem immense; few insects, indeed, possess a more powerful and sustained flight. It is often met with by ships at sea, to gain which it must have flown hundreds of miles from land. A specimen flew on board a steamer on her voyage from Africa, off Cape de Verde; and one has been taken by a fishing-boat in the North Sea, about a hundred miles east of May Island; and when *en route* for the Amazons, Wallace and Bates anchored off Salinas, at the pilot-station for vessels bound for

Para, they encountered two large hawk-moths. When pinched the moth seems able to exude an odour, which may be compared to that of jasmine or musk.

That it is Nomadic in Habit.

From this discovery of the moth at sea, frequently at a considerable distance from land, and in situations where it is impossible to suppose its presence to be involuntary, there seems no reason to doubt that it is nomadic in habit. It is an insect widely distributed. It is found over the whole of Europe, in Africa, and Western Asia, and in the eastern part of that continent is represented by a closely allied species; the genus, however, does not appear to occur in America. But while it has this wide geographical range, extending throughout Europe and Western Asia almost to the northern boundary of the colder temperate zone, probably its native home is in sub-tropical regions of India and Africa. In countries or districts within the colder temperate zone, as in Europe, its occurrence is of a somewhat fluctuating character; generally it is rather scarce, or only common in favourable years. In Great Britain, for instance, it has been found at some time or other in nearly every part of the country, from Land's End even to the Orkney and Shetland Isles;

but with one or two notable exceptions, of the south or south-eastern counties, where it is claimed to be observed almost every year, the records show its appearance to be uncertain. Roughly speaking, between 1865 and 1885 there was a period of a few years' duration when the species was common, followed by a long interval of scarcity; then the moth was common for two years in succession; after which, for a protracted time, in alternate years, it was generally scarce or locally common, followed again by an abnormal appearance and wide distribution. No doubt our entomological records are less full and complete than they might be, but from the observations we have, it is quite clear that the species is not of annual occurrence, but decidedly rare, or novel, to many localities. All the facts point to the Death's Head Moth being rather a wanderer in, than a denizen of, the greater portion of Britain, and that those which occur outside the one or two counties in which they may be said to be constant, are either migrants, or the offspring of migrants. May we not therefore believe that the rare specimens taken in the Orkneys, Shetlands, and such outlying portions of the kingdom, are visitors from the mainland (or the issue of these), who, in this case, obviously, have extended their rambles far from the place of their birth? Conceding nomadic proclivities to the moth, permits of us likewise easily accounting for its excessive abundance in certain years (otherwise a difficult problem to solve), by supposing that its numbers are at such times largely reinforced by immigrants from the Continent.

This subject of migration of Lepidoptera is beginning to attract the attention of entomologists, and it is expected that investigation will show that the abundance or scarcity of many species, besides the Death's Head Moth, is largely regulated by immigration. It remains for the future to disclose what are all the influences which cause the migration.

Somewhat unreliable in its times of appearance with us, usually the moth emerges in the autumn, in September or October—September appearing to be pre-eminently the month when it is on the wing—and probably hybernates, or else it lies in pupa through the winter; and hybernated specimens, or specimens from hybernated pupæ, are sometimes found in June. In a state of nature, in the opinion of some entomologists, the moths always emerge in the autumn, but when the pupa is kept under artificial conditions, the moth often appears in spring. Thus, Atropos seems taken from June to October inclusive, and it occasionally appears in November, though the latter date is exceptional. As for the larva, examples almost full-fed have been obtained at the end of June; the latest recorded date is October; but July and August, as already implied, appear to be the great months for this stage. It is these larvæ that, under favourable circumstances, produce imagines in September. The time of duration in pupa thus varies exceedingly, from a few weeks, as in the case just stated, to as many months, in the case of the late pupæ, whose appearance as perfect insects is delayed until the following spring or summer.

That we have but one generation of Atropos in the year is generally understood. Yet it strikes one, that larvæ found full-fed as early as June must surely attain the adult state before September. These larvæ have probably emerged from eggs laid by moths which on occasions are seen here in May, and, it may be suggested, they reach maturity by July or August, becoming in their turn the progenitors of the late September and October moths, and of the pupæ whose development remains in abeyance until the following year. Since hybernated pupæ in this country do not disclose the moths until about June, in all probability these May moths, the parents of the June larvæ, are immigrants.

On Rearing the Death's Head Moth.

The pupæ of Atropos, we have seen, are tender, sensitive subjects, and most difficult to rear. Putting things at their best, only about one in ten emerges. Under natural conditions, by means of the large chamber of the soil, and gummy secretion spun by the larva, the pupa evidently obtains freedom from irritation, and a more equable temperature and amount of moisture, and the reason of so many pupæ found not coming to maturity may lie in the very fact of this disturbance from their natural position. The difficulty of rearing is so great as regards those that do not turn to the moth in the autumn, that one of various plans for forcing them out is usually resorted to, as to keep them in a warm room, or even near a fire, always covered with moss, or like porous material, which is kept constantly damp; or they may be placed in bran, or fine sawdust. But probably, after all, in a general way, if simply protected from cold and left undisturbed, they will eventually yield the moths as well as if put through a course of "forcing."

On the other hand, it is a fact that the moth is not nearly so often met with as the larva. It may be, that while pupæ from British parentage require only protection from frost to bring them to maturity, pupæ which are the offspring of immigrants die unless aided by artificial warmth.

Due to the frequency of its death before completing its tranformations, indigenous* specimens of this remarkable and interesting moth are still deemed by collectors desirable acquisitions.

* Moths of immigrant parents are none the less British or indigenous. Wherever a moth naturally effects its metamorphoses, it must be recognized as belonging to the fauna of that country.

INDEX.

A

Acridian ears, situation of, 91
 different forms of, 91, 92
 forms in which they are absent, 92
 found in both sexes, and in most species, 92
 function difficult to determine, 92; possible solution of difficulty, 93
 in Stenobothrus, 92
 minute structure, 92
Acridiidæ, prominent characteristics, 82
 arranged in nine tribes, 132
 escape of the young from the egg, of *Stauronotus maroccanus*, 99; of Rocky Mountain Locust, 100
 some species present an unusual aspect, 132
Acridiides, 107, 108, 126, 133
Acrophylla titan, 46, 64
Aellopos titan, shot by Bates in mistake for a humming-bird, 222
Aëronaut, locust is an, 86
Affinities in nature, the series of, a concatenation or continuous series, 187
Aggressive Mimicry, 37
 Resemblances, and Protective, 25, 48, 52, 60, 67, 132, 136, 139, 153, 154, 183

Air-sacs, in Acridiidæ, 85
 absent in other Orthoptera, 86
 Acridiidæ remarkable amongst Orthoptera for, 85
 and powers of flight in locust, intimate association of, 86
 arrangement in Rocky Mountain Locust, 85
 found in various insects, 85
 how dilated, 86
 not found in larvæ, nor in truly apterous insects, 85
 use, 86
Alluring Colouration, 35
Ambulicinæ, 222
Ameles, prismatic capsules of, 20
Amphibious habits, strange, of some Tettigides, 134
Anabrus, increase to large numbers, 153
Anatomy of Acridiidæ, 83
Animula herrichii, strange larva case of, 209; *huebneri*, its larval case, 208
Anisomorpha, defence of, 49, 50
Anostostoma, the curious genus, 156
Ants as locust enemies, 130
Apterous Mantidæ and Phasmidæ compared, 24
Apterous Phasmidæ, 44, 46, 48, 59, 63

Apuleius, and the fable of Cupid and Psyche, 163, 174
Aschipasma, absence of elytra in, 43
Asilus flies, as locust enemies, 129
Ass, "the Golden," 163
Atrophy of wings of Mantidæ, 14

B

Bacillus, the genus, 76
Bacteria cornuta, eggs of, 53
Barber, Mrs., on "Voetgangers," 117
Bates, on *Aellopos titan*, its resemblance to a humming-bird, 222
 his encounter of hawk moths at sea, 238
 on the music of the Tananá, 148
Beauty of males of Psychidæ, probable cause of disappearance of, 214
Bee Hawk Moths, 222, 229
Bee-Robber, Death's Head Moth as a, 236
Bees, Death's Head Moth and, 237, 238
 that stridulous voice of Death's Head Moth controls, 237
Boisduval on beauty of the Uraniidæ, 194
 on division of Lepidoptera into Rhopalocera and Heterocera, 185
Brachystola magna, the "buffalo-hopper" or "lubber grasshopper," 140
Bradley, Richard, on the Walking-Leaf Insects, 72
Brongniart, on the name Eumegalodon, 159
 on post-embryonic development of *Schistocerca peregrina*, 101
 on Protophasmidæ, 78
Brough, Mr., on a species of Weta from Nelson, 158
Brunner, L., on colour difference of Oedipodides correlative with locality, 136
Buller, Sir W., on the "Wetas," 157

Butterflies, the most beautiful of all insects, 176
 favourite resorts, 176, 178
 interesting habits, 178-180
 like choice beds of flowers on the moist sand, 179
 not only gregarious, but migratory, 179
 of a quarrelsome disposition, 180
 peculiarities of highest interest in, 181
 twilight flyers, 180
Butterfly life, immense variety of, in equatorial zone, 176

C

Calasymbolus and Paonias, extremely beautiful genera, 228
Callidryas gathering in dense masses on ground, 179
Caloptenus spretus. See Rocky Mountain Locust
Capsule, egg, of *Mantis religiosa*, description of, 17
 consistency, 19
 explanation of manner of formation, 20
 situation of the eggs; the median chamber, 18
 subsidiary parts, 19
Capsules, egg, of Mantidæ, 16
Carabidæ as locust enemies, 129
Case Moths, in structure of female and in habit the strangest and most abnormal of all Lepidoptera, 204, 205
 females have become degenerate, 204, 214, 216, 217
Cases, larval, of Psychidæ, 205; enlargement, repair, locomotion, temporary suspension, complete withdrawal, mode of moving and retaining position within case, 209-211
 abundance of, 214; singular rarity of moths considering abundance of cases, 215, 216

Castnia, the genus, 187
  Fabricius on, 188
Castniidæ, an abnormal collection of pretty insects, 185, 187
  and Uraniidæ, 185
  geographical range, 190
  have most affinities with moths, 187
  in some respects combine the characters of both lepidopterous divisions, 187-190
Caudal horn of Sphingidæ, 220, 226
Cave-dwellers, 156
Cecidomyia, galls of, on willows, 143
  eggs of Locustidæ laid in, 143
Ceroys, a Peruvian, 65
  from Nicaragua, 65
Chæradodis, special Protective Resemblance of, 30, 31
  where found, 31
Chærocampa celerio, 229
Chærocampinæ, 222, 224, 225, 228, 229
  chiefly remarkable for power of swift and long-sustained flight, 223
Change of colour, in Phasmidæ, 44, 57, 102
  in Acridiidæ, 101
Characteristics and habits of Phasmidæ, 44
  Phasmidæ are herbivorous, 45
  Phasmidæ are sensitive to cold, 44
Chrysiridia madagascariensis, 200
Cladonotus humbertianus, 134
Cocytius cluentius, its long proboscis, 224
Colour changes in course of development in Orthoptera, 102
  of *Schistocerca peregrina*, 101
  of some Phasmidæ of the Phyllium group, 57, 102
Community of Descent, Darwinian doctrine of, a strong argument in favour of, 187
Conocephalides, 141, 159
  their head-ornament, 141
Coptopteryx females, 14, 24

Cry, singular, produced by Death's Head Moth, 233; theories to account for phenomenon, 233-235
Cupid and Psyche, fable of, 163, 174
Cursoria and Saltatoria, the series, 81

D

Dappula tertia, 213
Day-flying moths, 184
  migratory habits, 201
Death of Mantidæ, 17, 21
Death's Head Hawk Moth, 225, 230;
  larva, appearance, food, 231, 232;
  pupation, 232, 241-243
  an object of alarm to the superstitious, 235
  as a bee-robber, 236
  at sea, 238, 239, 241
  is sluggish, 238
  its cry, theories put forward to account for, 233, 234
  its grim feature, 231, 233, 236
  of superior dimensions, of nocturnal habit, 230, 231, 233, 235, 236, 238
  on rearing, 232, 242
  sound emitted by pupa, 234; by larva, its nature and cause, 234, 235
  that it is nomadic in habit, 239
  times of appearance, 232, 241, 242
Defence, means of, 25, 48, 49, 66, 67, 132, 136, 139, 153, 155, 157, 158
  aquatic habits, 50
  immobility, 45
  power of ejecting nauseous fluid, 49, 155
  prickles and spines, 49
Defence, positive, of Locustidæ, 154
Deinacrida, the curious genus, 156
  broughi, 158
  heteracantha, 156; its size, 157; food, and other habits, 157; clicking noise produced, 157
  megacephala, 157

Deinacrida thoracica, 157
Deroplatys, strange leaf-like appendages, 30, 31
Development of Mantidæ, 21
Devil's Riding-Horse. *See* Mantis
Diapheromera femorata, ravages of, 45
  eggs of, their deposition, 52, 54
  possesses gregarious tastes, 48
  post-embryonic development, 56
Digger-Wasps, as locust-enemies, 129
Dilatable tracheæ, 87
Dipterous parasites of Acridiidæ, 98, 130, 131
Disappearance of locusts from a spot invaded apparently inexplicable, 116
Dolichopoda palpata, a cave-dweller, 156
Dragon flies as locust enemies, 130

E

Ears, Acridian, 91
  of Locustidæ, 145
Egg-capsules of Mantidæ, 10
Egg-enemies of Acridiidæ, 97
  Anthomyia egg-parasite, 98
  birds and mammals, 98
  Cantharidæ, 97
  dipterous genus Idia, 98
  Locust-mite, 98
  Muscinæ, 98
  two-winged flies of family Bombyliidæ, 98
Egg-laying of Mantidæ, 16
  of Phasmidæ, 53
Egg-masses of Acridiidæ, details of, 95, 96
Eggs of Phasmidæ, remarkable nature of, 52
  each egg really a capsule containing an egg, 53
  manner of deposition, 53
  number produced, 54
  scramble out of, change during, 55
Eggs of Phasmidæ, their resemblance to seeds, 52; specially exemplified in the eggs of Phyllium, 54, 55
  time of laying, 54
Elephant Hawk Moths. *See* Chœrocampinæ
Ellema harrisii, its short tongue, 224
Emeralds. *See* Uraniidæ
Enemies of the locust, 97, 127, 129
  invertebrate, 97, 129; animals that feed upon the locust, or are parasitic externally, 129; animals that prey upon the locust internally, 130
  vertebrate, 98, 127; Locust Birds, 127; various vertebrates, 98, 128
Entometa ignobilis. *See* Lictor Moth
Ephippigerides, organs of stridulation, 147
Eremiaphila, first discovered by Savigny, 28
Eremobiens, 137, 140; modified to extraordinary extent for desert life, 137
Eugaster guyonii, its defensive fluid, 155
  its stridulation, unlike ordinary forms, 155
Eumegalodonidæ, 159
Extatosoma, an ugly monster, 49, 65

F

Fabricius on the genus Urania, 194
Females of Psychidæ, 204, 206, 214, 216
  change to imago, 206, 216; death, 206, 218
  perpetual prisoners, 206, 216
Fossil Phasmidæ, 78

G

Galls, deposition of eggs in, of *Meconema varium*, 143; of *Xiphidium ensiferum*, 143, 144

Geographical distribution of Mantidæ, 38; is very clearly defined, 39
"Golden Ass," the, 163
Gomphocerus, the genus, 140
Gongylus, 30, 35 : simulation to flowers, 37
Graber, on musical organs of females of Acridiidæ, 91
Graeffea coccophagus, ejection of fluid, 49 ; ravages of, 45
Green grasshoppers, 141
Ground species of Mantidæ, modifications of forms in general of, 28, 29
Eremiaphila, 28

H

Hair-streaks (Theclæ), 178
Haldmanella, 140
Hawk Moths. *See* Sphingidæ
  as wind visitors, 229
  highly specialized condition of some of the structural characters of peculiar interest, 220
  leading characteristics of this favourite group, 219
Head and sense organs of Mantidæ, 12, 13
Hemaris. *See* Bee Hawk Moths
  diffinis, 228
Henneguy, on the eggs of *Phyllium crurifolium*, 55
Hetaira esmeralda, a clear-wing butterfly, 177
Heteropteryx, its prickles and spines, 49, 65
Hymenopus bicornis, alluring colouring of, 36

I

Ichneumonidæ, no member of the family attacks locusts, 132

Immobility, a means of defence, 48, 49
  of Phasmidæ, reasons and use thereof, 45

K

Kallima, perfect Protective Resemblance of, 183
  inachis, 183
  paralekta, 183
Katydids, their music, 149, 154
  as pets, 150 ; results of confinement, 153
  have both a day and a night song, 150
  their pertinacity, 150
Künckel d'Herculais, on process of hatching of *Stauronotus maroccanus*, 99
  on dipterous parasites of *Stauronotus maroccanus* and other Acridiidæ, 98, 131
  on the Bombylid larvæ in the ova of *Stauronotus maroccanus*, 98

L

"Leaf-butterflies," 182
  Kallima inachis, 183
  Kallima paralekta, 183
"Leaf-insects," resemblance to leaves displayed by tegmina, by other parts, 67–69
  leaf-like tegmina possessed only by female, 70
Lefebvre, on the genus Eremiaphila, 28
Leg of an insect, typical development, 9
Legs, front, of Mantis, modification of in detail, 9
  principal function, 10
  secondary functions, 11
  the limb in repose, 11
  intermediate and posterior legs of Mantis, 11
Lepidoptera, 163, 176, 184, 203, 219, 230
  antennæ of, unsatisfactory as a classificatory basis, 185, 186

Lepidoptera, no one set of characters will serve as an infallible guide to distinguish moths from butterflies, 186
  Rhopalocera and Heterocera, 184, 185
  spine or spring on hind wings unsatisfactory as a classificatory basis, 186
Lictor Moth, its larval case, 208, 212; abundance of cases, 215
Locusta viridissima, its colour in assimilation with surroundings, 153
  musical organs, 147; stridulation, 148
Locustidæ prominent characters, 82, 141; absence of air-vesicles connected with tracheal system, peculiar head-ornament, ocelli generally imperfect, ovipositor, 141, 142
  chiefly nocturnal, 148
  defence of a positive nature, 154
  each species may ordinarily be distinguished by its peculiar note, 148
  food-habits, 153
  life-histories, 142; oviposition in earth, in plants, 142-144
  of a somewhat sedentary nature, 148
  Protective Resemblance, 153; perfection of resemblance of tegmina to leaves, 154
  readily distinguished from Acridiidæ, 82, 141
  remarkable forms, 156-159
  resemblance to Stick-insects, 155
Locusts, properly so-called, 106, 107
  their flight dependent on the wind, 86, 115, 116, 123, 124, 126
"Lubber grasshopper," the. *See Brachystola magna*

M

Macleay, that the minor natural groups of Lepidoptera often keep very constant to the same natural group of plants, 200

Macleay, on a species of Urania flitting about a grove of Pandanus, 200
Macroglossinæ, complete resemblance to humming-birds, 221
  placed at the head of the series Sphingidæ, 222
Manducinæ, 225
  Manduca atropos. *See* Death's Head Moth
Mantidæ, as locust-enemies, 130
  females compared with males, 14, 15
  striking characteristics, 9, 11
  voracity not limited to insects, 8
Mantis religiosa, devotional attitude, 3; meaning of habit, 6
  Brazilian name for, 5; not the saint but the tiger of the insect world, 5, 6
  geographical distribution, 40
  interesting mode of capture of prey, 6
  metamorphoses, 21
  name not indicative of pursuits, 5
  pugnacious propensities appreciated by Chinese, 8; and cannabalism, 8
  the subject of legends and superstitions, 3
Meconema varium, deposition of its eggs in galls, 143
Mecostethus, ears in, 92
Megalodon, the genus, 159
Megathymus yuccæ, the Yucca Borer, an interesting aberrant form, 190; regarded by some as a genuine butterfly, 190
  depredations for which famed, 190, 192
  funnel-like tube characteristic of larva, 192, 193
  habitat, appearance, habits, 190-192
  white powdery bloom, its use, 192, 193
Meso-thorax frequently of extraordinary length in Phasmidæ, 42
Metamorphoses of Mantidæ, 21
  development of organs of flight, 24
  interesting life and habits of the young, their moults, attitudes, food, etc., 22

Metamorphoses of Mantidæ, hatching of larva, time of, 21
number of larvæ yielded by a single capsule, 22
nymph state compared with sub-apterous or apterous adults, 24
time occupied in metamorphoses, 21
"Metamorphoses," the, or "Golden Ass," 163
Methone anderssoni, 137, 140
Metura elongata, a most interesting architect, 208
strange female of, 217
striking cases of, 208, 209. 215
Microcentrum retinerve, eggs of, 144;
metamorphoses, 144, 145
curious habits, 150
music, 149, 150
Migration of Butterflies, 179; of Day-flying Moths, 201; of Lepidoptera generally, 241
of Locusts, 109; laws governing, 111; phenomenon explained by excessive multiplication, 111, 113; other causes, remote, immediate, 113; remarkable manifestations of instinct attend, locusts take direction of predecessors, trial flights, locusts wait for change when wind unfavourable, 115; disappearance of locusts from a spot invaded apparently inexplicable, 116; distance to which swarms may migrate, 123; length of single flight, 123; facts in proof of power of prolonged flight, 124-126
Migratory, most species of Acridiidæ not, 106
disposition, not caused by anatomical differences, 108
locusts, habits and natural history, 109; ravaging power of, 109; huge size of swarms, 109; famine and pestilence result of a swarm, 110

Migratory species, exist in countries without giving rise to swarms, 108; penetrate to our shores, 108
species of Acridiidæ ascertained to be, of the Old World, 107; of the New World, 107
Mimetic resemblance of Phyllium, 67, 68
Mimicry of Phasmidæ, end gained by, 66
is purely defensive, 66
perhaps no other group in form and colour so generally imitative, 67
Morphidæ, 177
Murray, on Phyllium, 56, 58, 74
on the eggs of *Phyllium scythe* in Royal Botanic Garden, Edinburgh, 54
Music of Acridiidæ, 88
of Locustidæ, how produced, 147, 149
Musical organs of Locustidæ, situation, structure, 147
found usually only in male, 147

N

Nervous system, in Acridiidæ, 87, 88
the "brain," 87
Nomadic in habit, that Death's Head Moth is, 239

O

Oedipodides, 107, 126, 132, 136, 137
Eremobiens, 137
includes most of the species of migratory locusts of the Old World, 107, 136
winged, striking cases of colour difference, is correlative with locality, 136
Oiketicus kirbii, female imago of, 217
Orange river, attempted crossing of, by "Voetgangers," 121

INDEX.

Organization of Mantidæ, as a whole, 12
    is in conformity with a carnivorous life, 9, 12, 13
    is superior, 12
Organs of flight of Mantidæ, 13
Ornithoptera, the Bird-winged Butterflies, 177, 181
Oviposition in Acridiidæ, 94; gonapophyses, deposition of eggs, use of fluid discharged, soil preferred, time required for, extent of period, number of egg-masses, 95, 96
    in Mantidæ, 16
    in Phasmidæ, 53, 55

P

Pachytylus cinerascens, 107, 108
    marmoratus, 107
    migratorius, 107, 108
Packard, on arrangement of air-sacs in Rocky Mountain Locust, 85
Pages. *See* Uraniidæ
Pamphagides, 135
    Saussure, de, on their geographical distribution, 136
Paonias and Calasymbolus, extremely beautiful genera, 228
Papilio memnon, remarkable variety in the form of the females, 181, 182
Papilios in Nicaragua, sucking up moisture, 178
Parasites, of eggs of Mantidæ, 21
    of eggs of Acridiidæ, 98
Peringueyella, its slender stick-like forms, 155
Periodicity, no law of, governing destructive flights of locusts, 111
Phantasis, absence of elytra in, 43
Phasmidæ, general peculiarities, 41
    appearance grotesque, 41
    distribution of, 76
    enemies of, 48

Phasmidæ, marvellous imitative resemblance of vegetative objects, 60
    modification for aquatic life, 50
Phylliides, the tribe, 41, 68
Phyllium, moults of, 56–58
    number of species known, 68
    seasonal change of colour in, 57, 102
    specimens exhibited in Paris, 74
    the genus confined to tropics of Old World, 68
Plant Types of Mantidæ, 26
    their presentment of the phenomenon of Protective and Aggressive Resemblance, 25–27, 30, 35
Pneumorides, 132, 134
    Pneumora scutellaris, 135
Podacanthus, 47, 64, 76
Post-embryonic development, of Acridiidæ, 101; change of colour in course of development, 101, 102
    of Phasmidæ, 56; number of moults, 56
Poulton on disappearance of beauty of male Moths, 214
Prisopi, the, aquatic habits of, 50
Prochilus australis, resemblance to Phasmidæ, 155
Proscopides, 132
Protective and Aggressive Resemblance of Orthoptera, 25, 60, 67, 132, 136, 139, 153, 154
Prothorax of Mantidæ, modification of, 11
    its development shows its importance, 12
    its remarkable elongation and mobility, 11, 12; to what due, 12
Psychidæ, the Case Moths, 203
    change to imago, 206, 211, 216; supposed assistance of long sticks at event, 212; subsequent life of male, and general characters, 212, 213
    coming of age of male, preparation for pupal sleep, 211

Psychidæ, females are perpetual prisoners within larva habitaculum, 206, 216, 218
  larvæ subject to attacks of parasites, 215
  probable cause of disappearance of beauty in male, 214
  singular rarity of moths considering abundance of cases, 214, 215
  strangely difficult to rear, 215
  want of homogeneousness in perfect state of insects of this group, 213
Pterochroza, the genus, wonderful protective resemblance of, 154
Ptilota of Aristotle, the, 203

R

Rachitic condition of locusts, cause of, 130, 131
Reaumur, the earliest writer on sound made by Death's Head Moth, 233
Rearing Death's Head Moth, difficulty of, 232, 242
Riley, on hatching of Rocky Mountain Locust, 100
  on music of Locustidæ, 148, 149
  on music of *Microcentrum retinerve*, 149; on its interesting habits, 150
Rocky Mountain Locust, arrangement of air-sacs in, 85
  appearance, 108
  extent of period of oviposition, 96; number of egg-masses, 96
  migratory habit, 108
  process of hatching, 100

S

Sack-bearers, young, at home, 206
  ingenuity in construction of cases, 206, 208, 209

Sackträgers, 204
Sagides, some members of, 155
Saltatoria, term "grasshopper" applied to two families of, 82
Saussure, de, on slender stick-like forms in the genus Peringueyella, 155
  on oceans being impassable to locusts, 126
  on Oedipodides, 126
  on the geographical distribution of Pamphagides, 136
Savigny, on the genus Eremiaphila, 28
Schistocerca americana, 107
  peregrina, perhaps originally native to America, 126
  distribution, 107
  may deposit eggs at more than one spot during migration, 96
  occasionally penetrates to our shores, 108; crossing the ocean, 108, 126
  post-embryonic development, 101
Schizodactylus monstrosus, 158
Scorpions, as locust-enemies, 129
Scudder, on the song of Katydids, 150
  on the music of the Stenobothri, 90
  on the Protophasmidæ, 79
Sea, Death's Head Moth at, 238, 239, 241
Seas, that locusts traverse, of considerable width, 108, 125, 126
Sense-organs, in Acridiidæ, 83, 84, 91; sense of sight, of sound, of touch, and taste, 84, 92
  in Locustidæ, 141, 145
Sloane's, Sir Hans, history, on locusts crossing the ocean, 126
Smerinthinæ, 224, 225, 227, 228
Song, gift of, Acridiidæ remarkable for, 88
  apparatus for producing sound, 88, 89
  in aberrant forms of Acridiidæ, 88, 135, 137
  music characteristic of male, 90
  of importance to Acridiidæ, 88
Sound emitted by Death's Head Moth, 233, 236

Sound emitted by larva of, 234; its nature and cause, 234, 235
 by pupa of, 234
Special Protective and Aggressive Resemblance, beautiful examples of, 30, 35, 52, 60, 67, 132, 139, 154, 183
Spectre, the pink-winged, 64
Sphingidæ, their antennæ, 185, 219
 have their metropolis in the tropics, 228
 occasional visitors from sunny climes, 229
 pupation of, 220, 224, 225, 227
 sub-families of, 220; Macroglossinæ, Chœrocampinæ, Ambulicinæ, Sphinginæ, Manducinæ, Smerinthinæ, 220
 the natural position of the, 226
Sphinginæ, extraordinary length of tongue, 224
Sphinx-like attitude of larvæ of Sphingidæ, 220
Spiders as locust-enemies, 129
Spine, posterior, or caudal horn of larvæ of Sphingidæ, 220, 226
Stauronotus maroccanus, method of hatching, 99; the ampulla, 99, 100
 egg-parasites of, 98
Stenobothrus, music of, 89-91, 138, 140
 ears of, 92
Stridulating apparatus in Acridiidæ, 89, 135, 137; in the female, 90, 91, 138
Stridulation of Acridiidæ, 88, 135, 137
 during flight, 91
 of Locustidæ, 147
 specially characteristic of male, 90, 135, 138, 147
Superstitions, relating to Death's Head Moth, 235
 to Mantidæ, 3
Swainson on migration of Day-flying Moths, 201
Symbols of Psyche, 163, 176
 their conspicuous beauty and abundance, 176
Synemon, the genus, 190

T

Tachina-flies, as locust-enemies, 114, 130, 131
 annoyances of, as one of the causes of locust migration, 114
Tananá, its extraordinary music, 148; kept in cages for sake of its song, 148, 150
Tegmen of the female Phyllium, an exceptional structure, 70
Tegmina in Phasmidæ, usually of small size or absent, 42, 43
 attached to posterior part of mesothorax, 42, 43
Tettigides, 132-134
Tettix, the genus, 134, 140
Tibial ears, 145
 two principal kinds, 145; structure, 145, 146; function, 146
Tiger-beetles as locust-enemies, 129
Trachypetra bufo, Trimen on, 139
Trap-door spiders as locust-enemies, 129
Trimen on *Trachypetra bufo*, 139
Tropidoderus, 47, 64
Tryxalides, 133, 155
 resemblance to, of slender stick-like forms in the genus Peringueyella, 155
Twilight fliers, butterflies as, 180
Tyndall on sound, 94

U

Urania boisduvalii, one of the most beautiful Lepidoptera known, 195, 196
 egg-laying, larva, imago, 197-199
 inhabits Cuba, 195, 197, 199
 braziliensis, its migration, 200, 201
 fulgens, its migration, 200, 201
 leilus, 200
 sloanus, 200
Uraniidæ, the brilliant, 185, 194
 proved to belong to the Heterocera, 194

INDEX.

V

Vaal river, crossed by "Voetgangers," 118
"Voetgangers," interesting points in their natural history, 117
  on attaining maturity, 123
  their manner of travelling, 117, 118; no obstacle stays their course, how they cross rivers, their methods in water, 118, 121, 122

W

Walking-sticks proper, 41, 44–46, 48, 63; Wallace on, 63
  beautiful and giant-winged forms, 46, 64, 67
  bizarre shapes, 41, 65
  that they are a singularly isolated group, 79
  that they come of a remote antiquity, 78
Wallace, Dr., and Bates, their encounter of Hawk Moths at sea, 238

Wallace, Dr., on Alluring Colouration, 35
  on Walking-sticks proper, 63
"Weta-punga," the, 156
Wetas, the, an interesting group inhabiting New Zealand, 156–158
Wingless locusts, how they cross rivers, 118, 121, 122, 125
Wings, hind, in Phasmidæ, may be largely developed, 43, 46, 64, 67
  interesting provision for their defence, 43

X

Xerophyllum, the genus, 134
Xiphidium ensiferum, deposition of its eggs in galls, 143, 144
  time required for embryonic development, 144; for post-embryonic development, 144

Y

Yucca Borer, the, 190

THE END.

www.ingramcontent.com/pod-product-compliance
Lightning Source LLC
Chambersburg PA
CBHW031251250426
43672CB00029BA/2087